Muon and muonium chemistry

Muon and muonium chemistry

DAVID C. WALKER
Chemistry Department and TRIUMF, University of British Columbia
Vancouver, Canada

CAMBRIDGE UNIVERSITY PRESS
Cambridge
London New York New Rochelle
Melbourne Sydney

CAMBRIDGE UNIVERSITY PRESS
Cambridge, New York, Melbourne, Madrid, Cape Town, Singapore, São Paulo, Delhi

Cambridge University Press
The Edinburgh Building, Cambridge CB2 8RU, UK

Published in the United States of America by Cambridge University Press, New York

www.cambridge.org
Information on this title: www.cambridge.org/9780521103374

First published 1983
This digitally printed version 2009

A catalogue record for this publication is available from the British Library

Library of Congress Catalogue Card Number: 82-23636

ISBN 978-0-521-24241-7 hardback
ISBN 978-0-521-10337-4 paperback

CONTENTS

As with Wagner, it's not so bad as it sounds.

Mark Twain

PREFACE

The fast-breaking field of muon science is truly transdisciplinary. Using the tools of nuclear physics (as it examines the intrinsic decay of a short-lived, man-made, elementary particle one at a time) it draws inferences about normal chemical behaviour and about reactions which are not otherwise observable. Specifically, the muon is used to mimic the basic properties of our two best known elementary particles, the proton and the electron.

This book attempts to cover all aspects of the muon as a chemical entity, showing the current scope of the field, its major achievements and watersheds. Its main aim, however, is to try to make the muon zealots' poetic formalism intelligible and useful, through the medium of eponymous pedestrian prose, to chemists who deal with molecules composed of natural atoms.

My thanks are due to persons too numerous to list here. First, there are the authors whose data and ideas I have incorporated in this book. Then there are people whose work and effort provided the opportunity for my involvement with muons at TRIUMF: from the initial planners, designers and builders of this meson facility; to those who brought the muon spin rotation technique here; to the whole μSR group past and present; but particularly to my close associates in the research we are doing together - 'Jerry' Jean, Bill Ng, John Stadlbauer and Yasuo Ito. I am also most grateful to Anneke Rees who calmly and resourcefully transcribed to the disc of a word-processor my scribblings on thousands of paper scraps. Finally, to Gale and baby Elizabeth, who lived with those scraps of paper underfoot, and my preoccupation with them while we could have been doing other things together, in such good humour.

David C. Walker
Vancouver, August 1982

To G, S, J & e

1

INTRODUCTION

Although a muon exists for only two millionths of a second, this is time enough for it to participate in many of the fundamental chemical processes available to stable charged particles. One can produce either positive or negative muons, and since they have a mass intermediate between that of the proton and the electron, it is not surprising that the positive muon tends to behave as if it were a light proton, while the negative muon behaves as if it were a heavy electron.

1.1 Muons

Muons (μ) are elementary particles of the lepton family which occur, transiently, as part of the natural decay scheme of pions (π), as indicated by the overall sequence in Eq. [1.1].

$$\begin{array}{ccccc} \pi^{+/-} & \rightarrow & \mu^{+/-} & \rightarrow & e^{+/-} \\ (0.026\,\mu s) & & (2.2\,\mu s) & & (\infty) \end{array} \qquad [1.1]$$

Their origin and occurrence, therefore, require a source of pions. These are the lightest and commonest of the mesons, but they appear in nature extremely rarely, only when cosmic particles interact with the atmosphere. However, pions are readily created in one of three charge states (π^+, π^- or π°) during high-energy nuclear collisions. This is because nuclear glue can be thought of as the exchange of 'virtual' pions between nucleons, so that the production of pions outside the nucleus can occur when a nuclear collision exceeds the rest mass of a pion (0.15 amu, or 140 MeV). Consequently, copious fluxes of pions are manufactured at targets in the beams of the world's several accelerators capable of producing protons at energies much greater than 140 MeV. When the short-lived positive or negative pion decays it creates an energetic muon of the same charge. The muon lives for 2.2 μs, on average, before decaying to an energetic electron of the same charge. Table 1.1 gives some idea of the interrelation of these particles, and Figure 1.1 gives an indication of their production in a beamline of a high-energy proton accelerator.

During their 2.2-μs lifetime the muons can lose their kinetic energy by collisions with the molecules of a medium and thereby come to rest, and react chemically. In fact, this thermalization and initial chemical association takes a very short period of time ($<10^{-9}$ s in all media other than low pressure gases), so there is a relatively long period of time – from about 10^{-9} s to 2×10^{-6} s – during which the muon may be incorporated in a chemical entity which is capable of ordinary chemical behaviour, including diffusion and reaction. Once the muon decays, this chemical species disintegrates; but because of the muon's

Table 1.1. *Properties of the particles e, μ, π and p*

Symbol	Particle	Family	Mass (m_e)(a)	Spin	Antiparticle	Mean lifetime	Decay mode
e^-	Electron	Lepton	1	1/2	Positron (e^+)	∞	(e^+ annihilates)
μ^-	Muon	Lepton	207	1/2	Positive muon (μ^+)	2.2 μs	$\mu^- \rightarrow e^- + \bar{\nu}_e + \nu_\mu$
π^+	Pion	Meson	273	0	Negative pion (π^-)	26 ns	$\pi^+ \rightarrow \mu^+ + \nu_\mu$
p^+	Proton	Baryon	1836	1/2	Antiproton (p^-)	∞	(p^- annihilates)

(*a*) The mass of the electron, m_e, equals 0.000549 amu or 9.11×10^{-31} kg.

Figure 1.1. Schematic diagram showing the production of a μ^+ beam from a high-energy proton beam bombarding a Be target. Some of the π^+ produced are collected in the magnetic fields of a secondary beam-line, decay to μ^+, which then stop in the sample (S) and eventually decay to e^+.

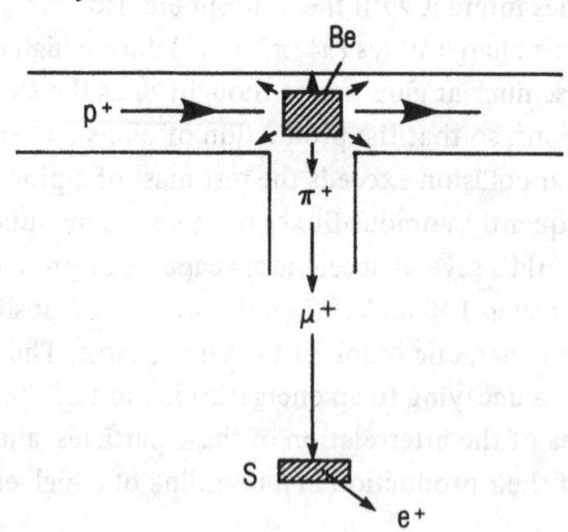

unique physical properties it is possible to follow its existence prior to decay and to make certain deductions about the chemistry that occurred.

The most useful attributes of the muon stem from its spin polarization, and from the violation of the parity invariance principle both as μ is formed and when it decays. This will be discussed in detail in chapter 3; but, for the moment, suffice it to say that there is a technique known as muon spin rotation (μSR) which enables one to observe the precession of the muon's spin in a transverse magnetic field, by observing the decay-electrons in a given direction as a function of time (Garwin, Lederman & Weinrich, 1957). The actual precession frequency depends upon the magnetic properties of the particular type of chemical association with which the muon is involved. In μSR one observes a spin-precession signal superposed on a lifetime histogram. This provides three pieces of useful information: from the frequency of precession one characterizes various chemical states of the muon; from the decay of the signal one evaluates a rate of reaction; and from the initial amplitude of the signal one calculates a yield - the probability that a muon will initially form that particular chemical state.

1.2 Muonium: a light isotope of hydrogen

In the days when only three elementary particles were known (the electron, proton and neutron) it was clear that hydrogen - specifically protium, ^{1}H - was the simplest possible atom. It also happens to be the most abundant atom in our universe, in the biosphere, in the human body, and in the living cell. But with the discovery of other charged particles, most of which are intrinsically unstable, it became possible to envisage, and to produce, other neutral 'atoms' which are probably more 'elementary' - at least in terms of quark theory. One such atom is muonium, used here with the chemical symbol Mu. It consists of the positive muon associated with an orbital electron. Muonium differs from protium only by virtue of having the light, short-lived positive muon as its nucleus rather than a proton. Its chemistry occupies nearly half of this monograph.

Although the mass of the muon is only one ninth that of the proton, it is still so much more massive than the electron that the reduced-mass of Mu is virtually the same as that of H. Thus the Born-Oppenheimer approximation can be used and one sees that the electron orbitals in H and Mu are equivalent in shape, size and energy. Therefore, muonium has essentially the same electronic structure, ionization energy, electron affinity, and Bohr radius as protium, deuterium and tritium. Their chemistries should differ from each other only by virtue of their different masses. In this sense Mu is unequivocally an *isotope* of H. (If it were argued that, by definition, isotopes must differ from each other only by the number of neutrons in the nucleus, then one should counter with the following argument. Within the framework of meson exchange theory, neutrons and protons are merely different charge states of the same nucleon: so that to 'extract' a neutron from a proton results in a pion (π^+) – which rapidly decays to a muon (μ^+)! Quark theory also, doubtless, allows one to circumvent the old definition of isotope in the case of muonium.)

With Mu as an isotope of ^1H, ^2H and ^3H, one has available the unique sensitivity provided by mass ratios of 9, 18 and 27. It is to be expected, therefore, that one of the most important roles for Mu in chemistry will be as a highly sensitive measure of mass-dependent effects, particularly, of kinetic isotope effects.

Some of the atomic properties of tritium, protium, muonium and positronium are given for comparison purposes in Table 1.2.

Muonium, like tritium, is radioactive. Hence it can be observed through its decay particle with great sensitivity, one at a time, using the powerful tools of nuclear physics. As already noted, for the specific case of the muon, it can also be studied through its spin polarization by μSR in addition to simple radiochemical techniques. One of the identifiable chemical states accessible to μSR is that of the free muonium atom (Hughes, McColm, Ziock & Prepost, 1960). Others include muonium incorporated in various free radicals. Furthermore, the yields and rates of reaction from these states are observed on the 10^{-9} to 10^{-5} s timescale – the very period over which many fundamental physicochemical interactions occur.

So that, in addition to its use in the study of isotope effects, muonium can act as a handle on hydrogen atom processes which are not observable by conventional methods. This use of muonium in monitoring H is its second major importance in chemistry. A third interest lies in studying muonium for its own sake – as an exotic artificial atom with some unique properties. The fourth interest lies in using Mu as a probe of the magnetic structure of materials.

Table 1.2. *Comparison of some of the properties of Mu with ^1H, ^3H and Ps (positronium, e^+e^-)*

Atom	Mass ($/m_e$)	Reduced-mass ($/m_e$)	Bohr radius (nm)	Ionization energy (eV)
^3H	5498	0.9998	0.05290	13.603
^1H	1837	0.9995	0.05292	13.599
Mu	208	0.9952	0.05315	13.541
Ps	2	0.5000	0.106	6.8

1.3 Other artificial atoms

Before delving into the chemistry of muons and muonium, it is appropriate to list other artificial 'atoms' derived from the combination of elementary particles, in case any of them is potentially superior to muonium as a light hydrogen isotope.

Table 1.3 lists the most plausible two-component neutral atoms under four categories: those with orbital electrons; their antiparticle equivalents; the particle-antiparticle pairs; and mesic atoms - with orbital muons or pions. Atoms containing kaons may be conspicuous by their absence from this list, in a few years time.

Although it has not yet been detected, π^+e^- ('pium') is probably formed when pions are stopped in many media. But, having zero spin (Table 1.1) there is no possibility of a pion spin rotation technique to match μSR. Its intrinsic lifetime will be given by that of the pion (26 nanoseconds), which is one hundred-fold shorter than Mu. Its chemical lifetime should be comparable to that of Mu in the same medium, except by virtue of its mass being 32% larger. From these considerations, π^+e^- seems to have no advantages over μ^+e^- as a hydrogen isotope, and has some distinct disadvantages.

None of the antimatter atoms has been produced in sufficient quantities for chemical study as they require an encounter between two rare particles (or an antimatter conversion). Likewise, for $\mu^+\mu^-$, $\pi^+\pi^-$ and p^+p^-, though the bound

Table 1.3. *Some of the simplest artificial atoms*

Atoms with electrons	Anti-atoms	Particle-antiparticle pairs	Mesic or muonic atoms
		e^+e^- (positronium)	
μ^+e^- (muonium)[(a)]	μ^-e^+	$\mu^+\mu^-$ ('mumuonium')[(b)]	$p^+\mu^-$ (muonic hydrogen)
π^+e^- ('pium')	π^-e^+	$\pi^+\pi^-$ ('pionium')	$p^+\pi^-$ (pionic hydrogen)
p^+e^- (protium)[(c)]	p^-e^+	p^+p^- (protonium)	
		$c\bar{c}$ (charmonium)	

[(a)] In retrospect one can see that the suffix 'onium' should really have been reserved for the particle-antiparticle pairs. However, the Mu atom which is the central theme of this book was named just after positronium. Following protium, deuterium and tritium, it should have been called muium; but since the positive muon is the antiparticle (because muons are leptons) it should be antimuium though muantium sounds better. It is probably too late to change now.

[(b)] This species should have been called muonium but will now have to be given an artificial name such as mumuonium. Hopefully pium and pionium will be used as in the Table.

[(c)] Protium is the only stable atom in the Table and, of course, occurs naturally as well as artificially.

states of each of these is important in particle physics. In addition, the bound states of charmonium (and other quarkoniums which are sufficiently heavy to be non-relativistic, such as bottomonium and toponium) have provided real advances in quantum chromodynamics with regard to colour force (Bloom & Feldman, 1982). The chemistry of the latter species could be of interest because of their very small size – but they are much more difficult to produce than the small mesic atoms. All of these species are rather short-lived for chemical studies. In fact positronium (Ps) is the only atom in columns 2 or 3 of Table 1.3 with any chemical significance at the present time. Ps readily occurs in most materials at the end of β^+ tracks, simply by the positron capturing an electron as it approaches thermal energy. Its chemistry has been widely studied, as referred to below.

Finally, the mesic atoms are of increasing importance in chemistry. There is, presumably, a whole Periodic Table of possible atoms in which an electron is replaced by μ^- or π^-. Muonic and pionic hydrogen merely represent the simplest such atoms. Due to the much larger reduced-masses of these atoms compared with protium, the μ^- or π^- orbitals are very much closer to the nucleus so these two atoms are extremely small. Transitions between orbitals involve high-energy X-rays, rather than visible or UV photons as in electron atoms. The existence of bound states of muonic atoms is proved by their characteristic X-rays. Lifetimes are typically of the order of microseconds and these atoms have an increasingly important role to play in certain branches of chemistry, including analytical chemistry (see chapter 10). Due to the strong interaction of mesons with nuclei, the analogous pionic atoms are much shorter-lived and the orbital transitions (X-ray production) are often severely curtailed by nucleon capture. Nevertheless, the chemistry of pionic hydrogen is under active study and of some general interest (Horvath, 1981).

But muonic or pionic hydrogen are of no importance as pseudo-isotopes of protium, because their Bohr radii and ionization energies are totally different from H. Also, their masses are virtually the same as H.

1.4 Positronium

When a positron associates with an electron in a bound quantum state it may appear either as a singlet ground state (*para*-Ps, with antiparallel spins) or as the triplet ground state (*ortho*-Ps, with parallel spins). These have virtually the same energy, but quite different lifetimes. The positron and electron of *p*-Ps annihilate by 2 γ emission with a mean lifetime of 0.125 ns. This is too short for *p*-Ps to be of much use in chemical studies, at least until the response time of γ-detectors and associated electronics improves considerably. On the other hand, *o*-Ps is restricted by symmetry rules to 3 γ annihilation and this is some

10^3-fold slower. Thus the intrinsic lifetime of *o*-Ps is some 140 ns; but this can be realized only *in vacuo*. In most condensed media, due to the density of electrons, the lifetime of *o*-Ps is reduced to 1 or 2 ns because of positron pick-off reactions and subsequent annihilation by 2 γ emission.

However, *o*-Ps lives just long enough in many systems of basic chemical interest for some of its fast chemical reactions to be studied through their effect on the *o*-Ps lifetime. It is also used as a probe of structure, particularly in solid state chemistry and physics. Despite real differences in the timescales and in the experimental methods involved, the development of muonium studies has been patterned to some extent on that of *o*-Ps. This will become evident from the references made to positronium in subsequent chapters.

But positronium should *not* be regarded as an isotope of hydrogen (or of muonium). In Ps there is no nucleus, because the two particles of which it is composed are of equal mass. As a result, they both orbit about the centre of mass and the effective size of Ps is about twice that of H. Thus the ionization energy is about half that of H, with the result that their basic chemical properties are quite different. As shown in Table 1.2, Ps and H are not a pair of atoms which differ merely in mass.

1.5 Isotope effects

Mass-dependent effects in chemistry mainly appear in three basic forms - kinetic effects, equilibrium (thermodynamic) effects, and spectroscopic or structural effects. It is the short lifetime of μ which militates against the use of μSR for equilibrium studies, and for most spectroscopy - except by the muonium-radical studies and by future possibilities for single photon spectroscopic techniques. On the other hand, it is this short lifetime which makes μSR feasible in the first place, and which allows one to probe the most interesting timescale of primary kinetic effects.

Kinetic isotope effects most often refer to the relative rate at which a reaction occurs when a reagent attacks a ^{1}H-X compared to a ^{2}H-X bond, for instance. Then the higher zero-point energy of the ^{1}H compound confers on it the faster rate, and one has a rate constant ratio favouring the lighter isotope. However, with regard to muonium studies, one is comparing the rate with which a free Mu atom reacts with a substrate relative to reaction by H atoms. Here, there are no differences in the zero point energies of the reactants, only of the activated complexes (transition states) and of the reaction products. Therefore, it is the heavier isotope which is favoured on energetic grounds.

It is unusual, but particularly useful, to have the isotope coming in the form of a free atom ready to react in a variety of ways (see section 1.6 below). By freeing the isotope from differences in vibrational and rotational energy in the reactants, one can focus attention exclusively on matters such as: quantum mechanical tunnelling; zero point energy effects in the activated complex; steric and orientation factors in the encounter; and the translational diffusion of the atom in solution - specifically its dependence on mass or volume.

Muonium rate data can be compared with ^1H data, where available, and with tritium and positronium. The detection techniques are totally different for these four atoms. ^1H data have largely been obtained by observing the ESR absorption as a function of time following pulse radiolysis in solutions in which H-atoms are produced - highly acidic water, for instance. The limitation here is on the speed of response of the ESR method and on materials sufficiently soluble and inert in the medium (Fessenden & Schuler, 1970). In typical tritium studies, one measures the stable products formed when recoil tritium is produced by a nuclear reaction in the presence of the substrate. One deduces the thermal reaction rates from competition studies, and one corrects for hot reactions through the effect of scavengers (Wolfgang, 1965). In positronium studies, one measures the reduction in the lifetime of *o*-Ps caused by the presence of a high concentration of substrate (Ache, 1979). One can also make inferences from angular correlation and Doppler broadening effects. So μSR studies, on the whole, are the most versatile, because one directly obtains many parameters: the yields, decay rates and magnetic properties of the different muon states; conversion between these states; the hyperfine coupling constants for muonium radicals; and electron spin-exchange cross-sections where appropriate (see chapter 8). It is also interesting to note that these four techniques make measurements on quite different timescales - Ps at 10^{-9} s, Mu at 10^{-9}–10^{-5} s, ^1H at 10^{-7}–1 s and ^3H at $>10^2$ s. Inferences, deductions or interpolations therefore have to be made into overlapping timescales in order to draw comparisons.

1.6 H-atom reactions

Hydrogen was doomed to be one of the most important atoms by being the simplest and commonest. But it is also one of the most versatile in its reactions. It is most frequently thought of as a 'reducing agent', as in Eq. [1.2]:

$$H + Fe^{3+} \rightarrow H^+ + Fe^{2+} \qquad [1.2]$$

but it may also act as a 'Bronsted acid' by transferring a proton, as in [1.3]

$$H + OH^- \rightarrow e^-_{aq} + H_2O \qquad [1.3]$$

or it can 'abstract', as in [1.4]

$$H + HCO_2^- \rightarrow H_2 + CO_2^- \qquad [1.4]$$

or 'substitute', as in [1.5]

$$H + CCl_4 \rightarrow CHCl_3 + Cl \qquad [1.5]$$

or 'add' to a multiple bond, as in [1.6]

$$H + CH_2{=}CHC_6H_5 \rightarrow CH_3{-}\dot{C}HC_6H_5 \quad [1.6]$$

or 'combine' with a paramagnetic molecule, as in [1.7]

$$H + O_2 \rightarrow HO_2 \quad [1.7]$$

or undergo 'electron spin exchange', such as [1.8].

$$H(\uparrow) + Ni^{2+}(\downarrow) \rightarrow H(\downarrow) + Ni^{2+}(\uparrow) \quad [1.8]$$

In addition to these, there are many other reactions accessible to H atoms when they are electronically excited or have excess kinetic energy ('hot'). Because of this amazing versatility, the kinetic isotope effects pertaining to free H atoms are of particular importance in chemistry. As chapters 7, 8 and 9 will show, kinetic isotope effects varying by four orders of magnitude have been found for Mu compared to H.

Summary

Muonium would seem to be ideal as a light isotope of hydrogen for kinetic studies and as a monitor of H-atom reactions. Luckily there is the valuable technique of μSR, originally developed for particle and nuclear physics but now used mainly for solid state physics and muonium chemistry. Unfortunately, one is limited to doing this work only at high energy accelerators producing intense fluxes of pions, which means there is merely a handful of centres* around the world where muon and muonium chemistry can be studied.

References to chapter 1

Ache, H. J. (1979). Positronium chemistry: present and future directions. In *Positronium and Muonium Chemistry*, Series **175**, ed. J. H. Ache, pp. 1–49. Washington: American Chemical Society.

Bloom, E. D. & Feldman, G. J. (1982). Quarkonium. *Scientific American*, **246** (5), 66–77.

Fessenden, R. W. & Schuler, R. H. (1970). Electron spin resonance spectra of radiation-produced radicals. In *Advances in Radiation Chemistry*, vol. **2**, ed. M. Burton & J. L. Magee, pp. 1–176. New York: Wiley-Interscience.

Garwin, R. L., Lederman, L. M. & Weinrich, M. (1957). Observations of the failure of parity and charge conjugation in meson decays: the magnetic moment of the free muon. *Physical Review*, **105**, 1415–17.

Horvath, D. (1981). Chemistry of pionic hydrogen atoms. *Radiochimica Acta*, **28**, 241–54.

Hughes, V. W., McColm, D. W., Ziock, K. & Prepost, R. (1960). Formation of muonium and observation of its Larmor precession. *Physical Review Letters*, **5**, 63–5.

Roduner, E. & Fischer, H. (1981). Muonium-substituted organic free radicals in liquids. Theory and analysis of μSR spectra. *Chemical Physics*, **54**, 261–76.

Wolfgang, R. (1965). The hot atom chemistry of gas-phase systems. In *Progress in Reaction Kinetics*, vol. **3**, ed. G. Porter, pp. 97–169. London: Pergamon Press.

*These include: SIN near Zurich, Switzerland, LAMPF in Los Alamos, USA, KEK in Tsukuba, Japan, JINR in Dubna, USSR, CERN in Geneva, Switzerland, and TRIUMF in Vancouver, Canada.

2

HISTORICAL PERSPECTIVES

The long-held belief that all the laws of physics would remain unchanged under space-inversion (parity), time-reversal, or charge-conjugation, was scuttled by events starting in 1956/57. The first to go was the principle of parity invariance, for processes governed by the weak interaction. It was found that neutrinos (ν) are always formed with negative helicity (spin and momentum vectors opposed), positive helicity for antineutrinos ($\bar{\nu}$). This means that the mirror image of any process involving either of these particles cannot occur, since mirror reflection (which is space-inversion, or parity) changes the sign of the helicity. For β^- particles formed in beta decay it means they all have the same helicity. For muons it means they are produced spin-polarized in reaction [2.1],

$$\pi^+ \rightarrow \mu^+ + \nu_\mu \qquad [2.1]$$

then decay asymmetrically (e^+ tends to be emitted along the muon's spin direction) in reaction [2.2].

$$\mu^+ \rightarrow e^+ + \nu_e + \bar{\nu}_\mu \qquad [2.2]$$

The muon spin rotation technique (μSR) utilizes both of these parity-violating interactions of the muon.

In the three-day period 15-17 January 1957, the editorial office of *The Physical Review* received three manuscripts: each one describing a separate novel experiment confirming the violation of parity in a weak interaction, as suggested by Lee & Yang (1956). The first of these papers described the experiments on polarized β^- decay that won for the principal author the Nobel Prize (Wu *et al.*, 1957). The second paper (Garwin, Lederman & Weinrich, 1957) verified parity-violation, by inventing the muon spin rotation technique and thereby proving the asymmetry inherent in reactions [2.1] and [2.2]. The third paper (Friedman & Telegdi, 1957) introduced the notion of a muonium atom, noting its spin-depolarizing effects due to the internal hyperfine interaction and the rapid precession of its total magnetic moment in the external magnetic field. These last two papers set the stage for μSR and established the principles by which most muon and muonium chemistry is studied.

The original apparatus of Garwin *et al.* (1957) is reproduced in Figure 2.1. Pions are stopped in the carbon absorber to give polarized muons, which trigger #2 and enter the target held in a variable magnetic field. Counters #3 and #4 act as a positron telescope, observing some of the positrons produced as the muons decay. The observed positron count rate was found to oscillate with a frequency which depended on the magnitude of the applied transverse magnetic field. This showed that the direction of positron emission was coupled to the precession of the muon's spin. These experiments utilized a fixed time interval with a varying magnetic field; but they are the precursors of present day μSR, in which one normally measures the positron count-rate as a function of time spent by the muon in the sample with a fixed magnetic field (see chapter 3).

By the following year, Swanson (1958) had shown that the observable muon polarization varied enormously, depending on the chemical composition of the target material. Most metals and liquid CCl_4 left the μ^+ polarization essentially undisturbed, but other solids and liquids showed a wide variation in the 'residual' quasifree μ^+ polarization (designated P_{res}). Following Friedman

Figure 2.1. Sketch of the apparatus originally used by Garwin, Lederman & Weinrich (1957) to verify the violation of the principle of parity invariance in the formation and decay of muons. It represents the first muon spin rotation (μSR) experiment.

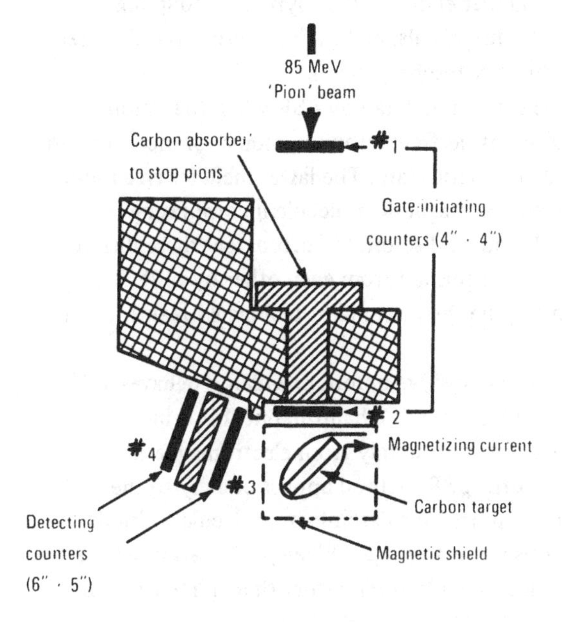

& Telegdi (1957), the missing polarization was attributed to the probable formation of muonium in some of these materials. This led, over a period of years, to the development of reaction models and theoretical treatments connecting the observed P_{res} to the mechanism of muon depolarization (Nosov & Yakovleva, 1965; Firsov & Byakov, 1965; Ivanter & Smilga, 1969; and Brewer *et al.*, 1971).

However, the formation of a free muonium atom (Mu) was not merely speculative because Hughes, McColm, Ziock & Prepost (1960) had reported its direct observation in argon gas, in 1960. Furthermore, they quickly recognized the potential role of Mu in chemistry as a light isotope of hydrogen and as a convenient handle on certain types of H-atom reactions (Hughes, 1966). Some of its chemical reaction rate constants were then measured, by adding known amounts of substrates to the argon gas in which muonium was formed and determining its rate of disappearance (Mobley *et al.*, 1966, 1967; Mobley, 1967).

Although the effect of solutes on the residual quasifree μ^+ polarization was used to infer the presence and reaction rates of muonium in liquids (Brewer, Gygax & Fleming, 1973; Brewer *et al.*, 1974), it was not until 1976 that muonium was observed directly in a liquid through its characteristic precession. This was first achieved using pure water (Percival *et al.*, 1976) but Mu has since been observed in alcohols (Percival, Roduner & Fischer, 1978) and saturated hydrocarbons (Ito, Ng, Jean & Walker, 1980).

The next major step in muon chemistry was the discovery of muonium-containing free radicals, in the μSR spectrum of unsaturated hydrocarbons at relatively high magnetic fields (Roduner *et al.*, 1978). Hyperfine coupling constants were measured for these Mu-radicals, and their structure was deduced by analogy with H-radicals (Roduner & Fischer, 1981).

At the present time, μ^+ can be studied in detail by μSR when the muon is present in one of *three* chemical states: as free muonium atoms; or incorporated in free radicals; or present in a diamagnetic state. The latter includes free muons, solvated or trapped muons, and any diamagnetic molecule incorporating the muon, such as MuH, MuOH, $C_6H_{13}Mu$, and so on. Unfortunately these different diamagnetic states cannot yet be distinguished from each other by μSR. Perhaps achieving such an identification may be the next major technical breakthrough for muon chemistry.

All of the above has referred to the positive muon, in which μ behaves as if it were a light proton. In fact it is a lepton, not a hadron, but this is inconsequential as the chemical states discussed arise only from electromagnetic interactions (Coulomb forces). But the μSR method applies equally to the negative muon. Whereas μ^+ is not captured into atomic or molecular orbitals and has no nuclear interaction, this is not so for μ^-. When μ^- is used it behaves as if it were a heavy electron, taking on an 'orbital' rather than a 'nuclear' role.

Its chemistry is totally different. At the moment its study is of much more limited scope than that of μ^+, but its potential will be discussed briefly in chapter 10.

Finally, a historical note about the muon as an elementary particle. Although most chemists are content to regard matter as being composed only of e^-, p^+ and n, it is fifty years since this was known to be an approximation. Our muon was the fifth particle to be discovered - originally mistaken for Yukawa's meson when first detected in cosmic rays in 1937. It has survived the rise to ~36, the fall back to ~12 and the rise again to ~48 in the number of elementary (funda-

Figure 2.2. This plot shows, approximately, how the number of indivisible particles has varied over the years during this century. The muon was the fifth to be discovered and has survived the turmoil. In some cases the year refers to experimental discovery in other cases to theoretical deductions.

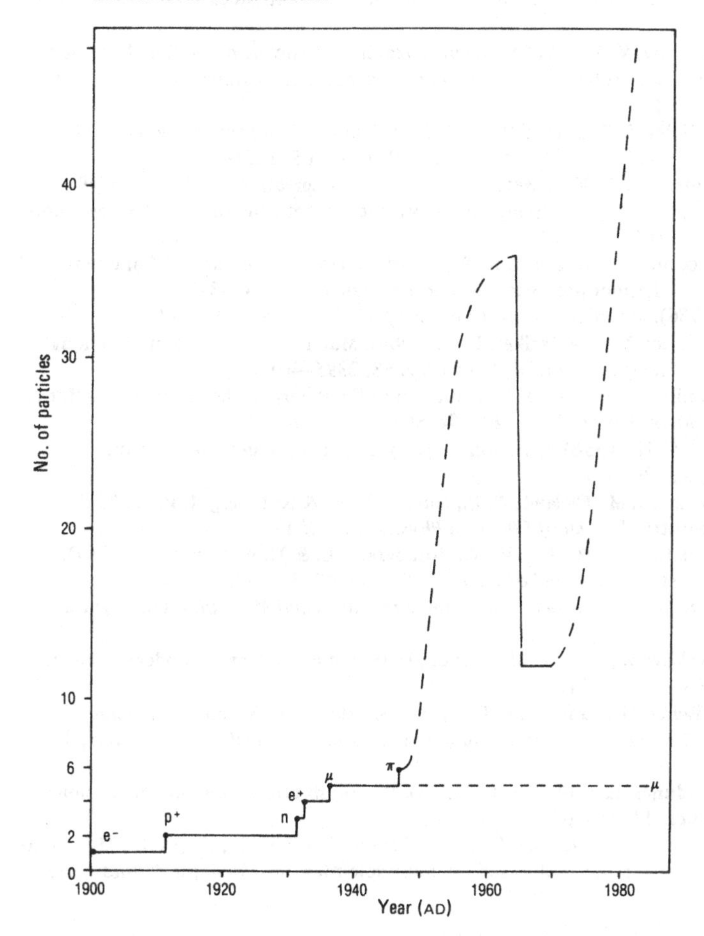

mental) particles as indicated in Figure 2.2. And it is interesting that the proton – with which we are mainly comparing μ^+ – did *not* survive the fall of 1965: for it is now thought to be composed of three quarks and to have a finite lifetime (Weinberg, 1981). Whether that makes Mu more 'elementary' than H depends on one's perspective; but it adds extra credence to the pursuit of muon and muonium chemistry.

References to chapter 2

Brewer, J. H., Crowe, K. M., Johnson, R. F., Schenck, A. & Williams, R. W. (1971). Fast depolarization of positive muons in solution – the chemistry of atomic muonium. *Physical Review Letters*, **27**, 297–300.

Brewer, J. H., Gygax, F. N. & Fleming, D. G. (1973). Mechanism for μ^+ depolarization in liquids – muonium chemistry with radical formation. *Physical Review*, **A8**, 77–86.

Brewer, J. H., Crowe, K. M., Gygax, F. N., Johnson, R. F., Fleming, D. G. & Schenck, A. (1974). Muonium chemistry in liquids: evidence for transient radicals. *Physical Review*, **A9**, 495–507.

Firsov, V. G. & Byakov, V. M. (1965). Chemical reactions involving muonium. A method for determining the absolute rate constant and other reaction parameters. *Soviet Physics* JETP, **20**, 719–925.

Friedman, J. I. & Telegdi, V. L. (1957). Nuclear emulsion evidence for parity nonconservation in the decay chain $\pi^+ \rightarrow \mu^+ \rightarrow e^+$. *Physical Review*, **105**, 1681–2.

Garwin, R. L., Lederman, L. M. & Weinrich, M. (1957). Observations of the failure of parity and charge conjugation in meson decays: the magnetic moment of the free muon. *Physical Review*, **105**, 1415–17.

Hughes, V. W., McColm, D. W., Ziock, K. & Prepost, R. (1960). Formation of muonium and observation of its Larmor precession. *Physical Review Letters*, **5**, 63–5.

Hughes, V. W. (1966). Muonium. *Annual Review of Nuclear Science*, **16**, 445–70.

Ito, Y., Ng, B. W., Jean, Y.-C. & Walker, D. C. (1980). Muonium atoms observed in liquid hydrocarbons. *Canadian Journal of Chemistry*, **58**, 2395–401.

Ivanter, I. G. & Smilga, V. P. (1969). Contribution to the theory of the chemical reactions of muonium. *Soviet Physics* JETP, **28**, 796–801.

Lee, T. D. & Yang, C. N. (1956). Question of parity conservation in weak interactions. *Physical Review*, **104**, 254–8.

Mobley, R. M., Bailey, J. M., Cleland, W. E., Hughes, V. W. & Rothberg, J. E. (1966). Muonium chemistry. *Journal of Chemical Physics*, **44**, 4354–5.

Mobley, R. M., Amato, J. J., Hughes, V. W., Rothberg, J. E. & Thompson, P. A. (1967). Muonium chemistry, II. *Journal of Chemical Physics*, **47**, 3074–5.

Mobley, R. M. (1967). *Interaction of Muonium with Atoms and Molecules*. Ph.D. Thesis, Yale University.

Nosov, V. G. & Yakovleva, I. V. (1965). Depolarization of μ^+ mesons in condensed media. *Nuclear Physics*, **68**, 609–31.

Percival, P. W., Fischer, H., Camani, M., Gygax, F. N., Ruegg, W., Schenck, A., Schilling, H. & Graf, H. (1976). The detection of muonium in water. *Chemical Physics Letters*, **39**, 333–5.

Percival, P. W., Roduner, E. & Fischer, H. (1978). Radiolysis effects in muonium chemistry. *Chemical Physics*, **32**, 353–67.

Roduner, E., Percival, P. W., Fleming, D. G., Hochmann, J. & Fischer, H. (1978). Muonium-substituted transient radicals observed by muon spin rotation. *Chemical Physics Letters*, **57**, 37–40.

Roduner, E. & Fischer, H. (1981). Muonium-substituted organic free radicals in liquids. Theory and analysis of μSR spectra. *Chemical Physics*, **54**, 261–76.
Swanson, R. A. (1958). Depolarization of positive muons in condensed matter. *Physical Review*, **112**, 580–6.
Weinburg, S. (1981). The decay of the proton. *Scientific American*, **244** (6), 52–63.
Wu, C. S., Ambler, E., Hayward, R. W., Hoppes, D. D. & Hudson, R. P. (1957). Experimental test of parity conservation in beta decay. *Physical Review*, **105**, 1413–15.

3

μSR TECHNIQUES

It is the parity-violating emission of a positron preferentially along the instantaneous spin direction when the muon decays, that makes μSR possible. The muon decays independently of its environment, of course; but prior to decay, its spin would have been coupled to unpaired electrons or nearby nuclei and it would have undergone Larmor precession in magnetic fields transverse to the spin direction. Therefore, some measure of the chemical state occupied by the muon during its short lifetime is revealed by its spin vector at the moment of decay.

There are three distinct types of techniques each utilizing the muon's asymmetric decay and its spin polarization, which happen to be covered by the acronym μSR: *m*uon *s*pin *r*otation, muon spin *r*elaxation, and muon spin *r*esonance. These different methods are illustrated in Figure 3.1. In the first, one measures the rotation of the muon spin in a transverse magnetic field; in the second, one follows the relaxation of the initial spin polarization in a longitudinal magnetic field; and in the third, one observes the polarization through the resonant absorption of microwave power due to transitions between hyperfine substates. Most of the recent chemical studies have been performed using the first – the rotation method – so the major emphasis here will be on it. Basically it is used in three ranges of magnetic fields: 2-15 G (1 gauss = 10^{-4} tesla, T) for free muonium atoms (MSR); 50-200 G for diamagnetic muon states (μSR); and 500-5000 G for Mu-containing free radicals (MRSR). Most chemical μSR is based on the effects of externally applied and controlled magnetic fields; but a lot of solid state physics utilizes μSR to investigate internal fields, particularly to probe the local magnetic structure of the host medium in which μ^+, μ^- or Mu reside. Furthermore, μSR can be used in d.c. or pulsed modes, but most chemical studies thus far have been performed under d.c. conditions.

μSR applies equally well to μ^+ and μ^-, since both show the same parity-violating asymmetries in their formation and decay, Eqs. [3.1-3.4].

$$\pi^+ \rightarrow \mu^+ + \nu_\mu \qquad [3.1]$$

$$\pi^- \rightarrow \mu^- + \bar{\nu}_\mu \qquad [3.2]$$

$$\mu^+ \rightarrow e^+ + \nu_e + \bar{\nu}_\mu \qquad [3.3]$$

$$\mu^- \rightarrow e^- + \bar{\nu}_e + \nu_\mu \qquad [3.4]$$

However, again mainly from a chemical point of view, μ^-SR is not as interesting or important as μ^+SR because antimuonium (μ^-e^+) is exceptionally rare, and the main interest in μ^- lies in its ability to form muonic atoms (chapter 10). Only μ^+SR is considered in any detail here.

3.1 Muon production and decay

High fluxes of muons are most easily produced from the decay of pions generated by nuclear reactions using high-energy proton beams. Details vary greatly from one centre to another. Typically, however, a 300–1000 MeV proton

Figure 3.1. Representation of the three types of μSR techniques. (a) Muon spin rotation (the field H is perpendicular to the muon spin); (b) muon spin relaxation (field H_0 parallel to spin: (Forward (FWD) count rate higher than Backward (BWD)); (c) muon spin resonance (microwave or rf field applied to the target in addition to the longitudinal magnetic field). The heavy arrow represents the initial muon spin, N is the count rate in time-bin at time t, ω_L is the Larmor frequency.

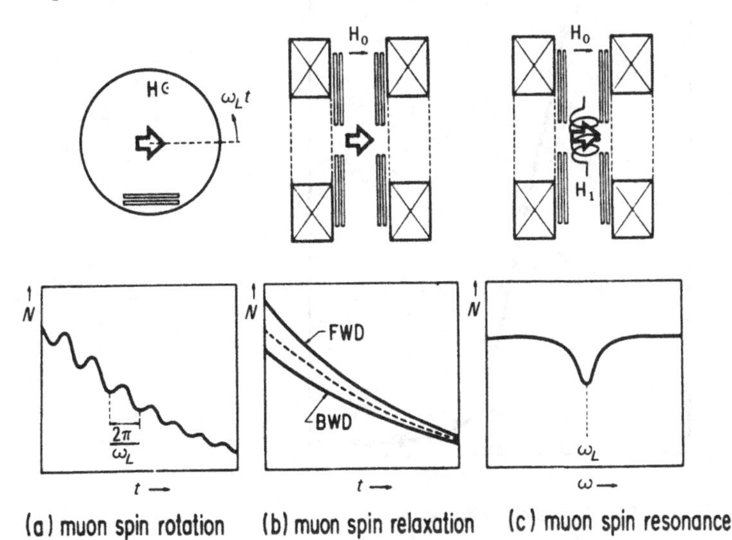

beam is made to strike a beryllium target to produce π^+ by nuclear reactions such as $^9Be(p^+,\pi^+)^{10}Be$. Pions have zero spin and decay to muons with a mean lifetime of 26 ns in accordance with Eq. [3.1]. The unique helicity of the muon's neutrino (ν_μ) confers a definite (negative) helicity (spin opposite to linear momentum) on the muon in the pion's rest frame. This ensures that the muons coming from pions which are 'at rest' will be 100% longitudinally-polarized with spins opposite to their direction of motion. They are also essentially mono-energetic.

Beams of muons originating from pions which come to rest in the outer skin of the Be production-target all have an energy of 4.1 MeV (corresponding to the muon momentum as a result of the formation Eq. [3.1], 29.8 MeV/*c*). This confers on them a total range of only ~0.13 g cm^{-2} in low *Z* materials (1.3 mm in water, for instance) as shown in Figure 3.2. Although the weak penetration presents problems for beam windows and front counters, this 'Arizona' or 'surface' beam (Pifer, Bowen & Kendal, 1976) is particularly well suited to the

Figure 3.2. The range of 'surface' muons is shown as the fraction of beam intensity remaining after passing through various thicknesses of mylar (converted through density to mg/cm^2). Some degradation of the initial 4.1 MeV energy had occurred due to transmission through the thin beamline window. Data taken from Brewer & Crowe (1978).

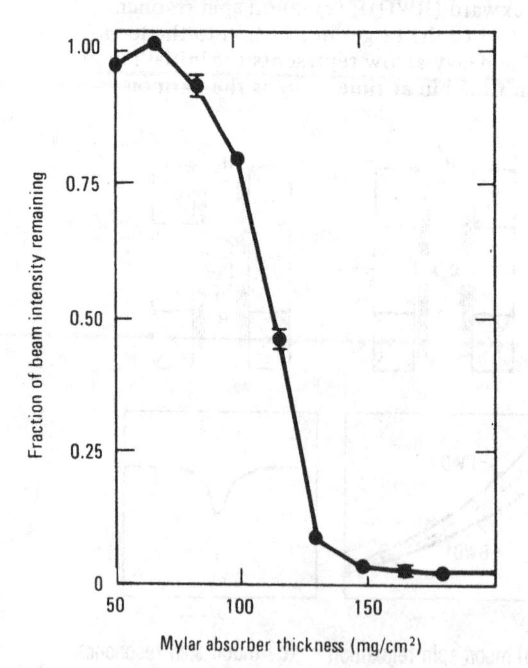

study of gases at one atmosphere pressure, very thin or rare solids, or small quantities of liquids held in thin-walled cells.

Alternatively, the pion is allowed to decay 'in flight' within a beamline and the muons are collected from the forward or backward direction of the pion's motion. The pion's momentum influences the muon's initial momentum and in some cases its helicity. The beam of 'forward' muons typically is of high intensity – though badly contaminated with p^+, π^+ and e^+ of the same momentum – of relatively high energy, with negative helicity and polarization of 50–80%. By contrast, the 'backward' muon beam is of much lower flux, but very much cleaner, of positive helicity and 60–80% polarized. These muons are sufficiently energetic to penetrate thick-walled glass cells and are not seriously bent by transverse magnetic fields up to 5 kG. Most of the chemical data to be discussed in this monograph were obtained with either 'surface' or 'backward' muon beams. For rare experiments it could be useful to be able to employ muons with either, or both, helicities.

The energetic muons are made to bombard the sample under study and to stop within it. During their thermalization, or shortly thereafter, the muons interact with the medium in a variety of possible ways depending on its properties. This is the subject of chapter 6; but, basically, the muon exists between $\sim 10^{-9}$ and 10^{-6} s after injection into the medium in one of three magnetic states: as free muonium atoms (μ^+e^-, Mu); in Mu-containing free radicals; or in purely diamagnetic associations. It will be convenient to use the subscripts D and M to refer to parameters of the diamagnetic states and muonium respectively. Some of the physical parameters of bare muons and muonium atoms are provided in Tables 3.1 and 3.2.

These three types of magnetic states of the muon are readily distinguishable by muon spin rotation, even when present together in a medium. The technique utilizes an external magnetic field transverse to the muon's spin, so that the latter precesses in the plane normal to the field, at a rate set by the effective magnetic moment of the muon in its particular chemical state. The actual direction of the spin at the moment when the muon decays thus depends on

Table 3.1. *Some physical parameters of positive muons*

Spin = $\frac{1}{2}$
Lifetime = 2.1971×10^{-6} s
Mass = 0.1134 amu = 206.77 m_e = 0.1126 m_p
Magnetic moment = 4.49048×10^{-30} J G^{-1}
3.18334 μ_p = 0.004836 μ_e
g-factor = 2.0023318 = 1.000006 g_e
Gyromagnetic ratio = 13.5544 kHz G^{-1}

three factors: the initial direction (all the same, since a spin-polarized beam is used); the rate of the precession (characteristic of the chemical state and the magnetic field); and the lifetime of the particular muon (i.e. the time it has been in that chemical state).

One calls on the inherent asymmetry in Eq. [3.3] in order to see this change of spin direction during the lifetime of the muon. Since the positron is emitted preferentially along the muon's spin, one merely needs to determine that direction, and the lifetime, for each muon.

3.2 Muon spin rotation (μSR)

In the experiments which pioneered the μSR technique, Garwin, Lederman & Weinrich (1957) positioned a positron 'telescope' in the plane containing the polarization axis and gated the observation time (1.25 μs) at a fixed time (0.75 μs) after arrival of an incident muon. They observed the asymmetry by varying the precession frequency (ω), through alteration of the magnetic field ($\omega = geH/2mc$). Very shortly thereafter, Swanson (1958) reported data obtained through the more efficient utilization of a limited muon flux, thereby originating the method that is mainly used today. He fixed the precession frequency with a constant field, then utilized the full muon lifetime spread by electronically measuring the lifetime of each muon whose spin was observed in a given direction.

The basic technique can be illustrated with reference to the sketch in Figure 3.3. An incident muon passes through the front scintillation counter (T) triggering a 'start' signal to a clock. The muon enters and stops in the medium, associates chemically, and its spin starts to precess in the transverse magnetic field. Some 2.2 μs later, on average, the muon decays, with the energetic positron (up to 52.8 MeV) emitted anisotropically. If this happens to be directed towards the positron telescope, consisting of three scintillation counters R_1, R_2 and R_3 arranged in 'coincidence' to minimize background events, then a 'stop' signal will be sent to the clock. The time delay between 'start' and 'stop'

Table 3.2. *Some physical parameters of the muonium atom*

Spin states: ^{T}Mu ($S=1$) and ^{S}Mu ($S=0$)
Mass = 0.1140 amu = 207.8 m_e = 0.1131 m_p
Reduced mass = 0.9956 times that of H
Bohr radius = 1.0044 times that of H
Ionization energy = 0.9956 times that of H
Mean velocity at 300 K = 7.5×10^3 m s^{-1} = 2.97 times that of H
deBroglie λ (300 K) = 2.98×10^{-10} m = 2.97 times that of H
Hyperfine frequency (ω_0) = 4463 MHz (2.8044×10^{10} rad s^{-1})

represents one data point in a lifetime histogram. This delay time is handled through a time-to-amplitude converter and binned in a counting device such as a multichannel analyser or a computer. Only a small fraction of the total number of emitted positrons are seen by R_1-R_3. If no stop signal is received within 8 μs after a muon start, then the clock is reactivated to await a new muon. This allows a muon flux up to $\sim 10^5 \, s^{-1}$: beyond this the individual μ^+-e^+ correlation is lost. The process is repeated until $\sim 10^7$ good events are collected for a histogram to be constructed.

Not all positrons are emitted with the maximum possible energy, or exactly along the muon spin direction. Instead there is a broad distribution of energies and angles subtended (θ) between the muon spin and the direction of positron emission. In fact the anisotropy is quite limited. Weak interaction theory

Figure 3.3. Schematic arrangement of the basic μSR apparatus. Muons (μ^+) pass through the triggering 'start' counter T and stop in the sample S which is held in a transverse magnetic field created by Helmholtz coils H. Some of the positrons (e^+) formed by decay of μ^+ travel to the counter telescope R_1–R_2–R_3 to provide a 'stop' signal. In practice two or more independent e^+ telescopes are used to collect the positrons emitted in a few directions in the plane of precession. Absorbers are often placed between the counters of a telescope to reduce the low-energy (low polarization) positrons in order to increase the mean e^+ energy and, therefore, the asymmetry (A_0 in Eq. [3.5]).

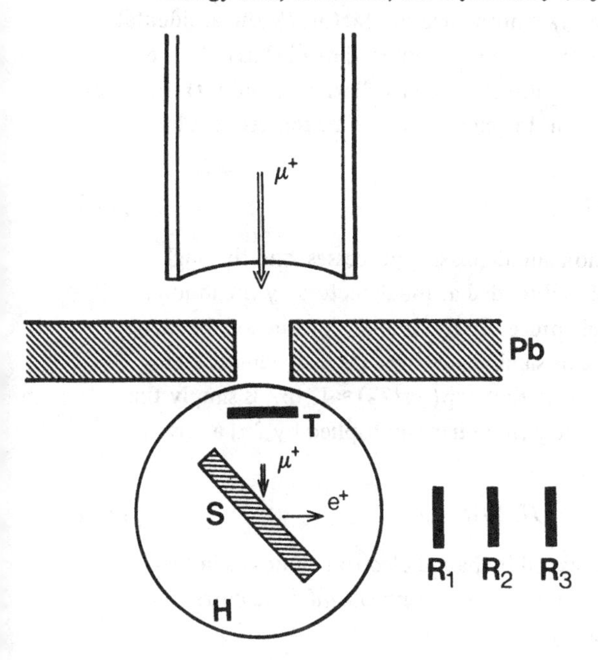

predicts that the three-body decay of μ^+ (Eq. [3.3]) gives a positron emission probability proportional to $(1+A\cos\theta)$, where A would equal 0.33 of the full polarization, but for several experimental imperfections (Brewer, Crowe, Gygax & Schenck, 1975) including: the finite solid angle of the detectors, and all positron energies not being counted with equal efficiency. Therefore, even a positron telescope which is at 90° to the incident muon spin direction, as in Figure 3.3, detects a significant fraction of decaying muons without precession. With zero magnetic field all positron counters simply give a smooth lifetime histogram, providing there is no depolarization.

However, when a transverse field is applied, the probability of receiving a positron signal increases as the spin precesses towards the telescope, then falls to a minimum as the spin reaches the opposite direction. Therefore the precession is seen as an oscillation superposed on the lifetime histogram. The frequency of the oscillation depends on the magnetic state of the muon and its magnitude is related to the number of muons in that state.

Figure 3.4(a) shows typical raw μSR data – taken for the particularly simple case of CCl_4 where all muons form diamagnetic states. This μSR 'spectrum' gives the counts in a lifetime bin plotted against that lifetime. As shown in Figure 3.4(b) the data points can be fitted by the computer to an equation such as [3.5],

$$N(t) = N_0\{Bg + \exp(-t/\tau)[1+A_0 f(t)]\} \qquad [3.5]$$

where $N(t)$ is the bin count-rate, N_0 a normalization factor, Bg the accidental time-independent background, τ the muon's mean lifetime (2.2 μs), A_0 the normalized initial 'asymmetry' (amplitude of the oscillations), and $f(t)$ the time-dependence of the spin-polarization. In general $f(t)$ can be represented by Eq. [3.6],

$$f(t) = \exp(-t/T_2)\cos\theta \qquad [3.6]$$

where T_2 covers all spin relaxation and dephasing processes, and the angle θ equals the initial phase ϕ_D (angle subtended at the detectors by the muon's initial polarization) plus the angle precessed by time t at precession frequency ω_D. T_2 is usually too large to be of significance because diamagnetic states of muons are generally unreactive, so that $\exp(-t/T_2) \approx 1$. ω_D is simply the nuclear Larmor precession frequency (in radians multiplied by 2π) as given by Eq. [3.7]

$$\omega_D = g_\mu eH/2m_\mu c = 13.55H \text{ kHz} \qquad [3.7]$$

for bare muons and for those screened by paired electron clouds as in diamagnetic molecules. (As noted earlier, simple chemical-shifts are currently beyond μSR frequency resolution.)

3.3 Muonium spin rotation (MSR)

Muonium is a two spin-1/2 system existing as 'singlet' or 'triplet' states, depending whether the muon and electron spins are paired ($S=0$, $|\alpha_\mu \beta_e\rangle$) or unpaired ($S=1$, $|\alpha_\mu \alpha_e\rangle$). For convenience these will sometimes be referred to

Figure 3.4. Typical μSR histograms presented in two ways: both are for samples of CCl_4 in 100-G transverse fields. Upper, the raw data in the form of the observed count-rate as a function of muon lifetime (trace connects data from 10-ns bins). The mean muon lifetime is indicated at 2.2 μs. Data from Brewer *et al.* (1975). Lower, the data points (every tenth with an error bar) after fitting by a computer with the time-independent background (<10%) removed, and the exponential fall-off in the number of muons remaining (e^{-t}/τ) removed. The line is the computer's χ^2 minimization plot of the equation $A(t) = A_D \cos(\omega_D t + \phi_D)$. Data from Ng (1980).

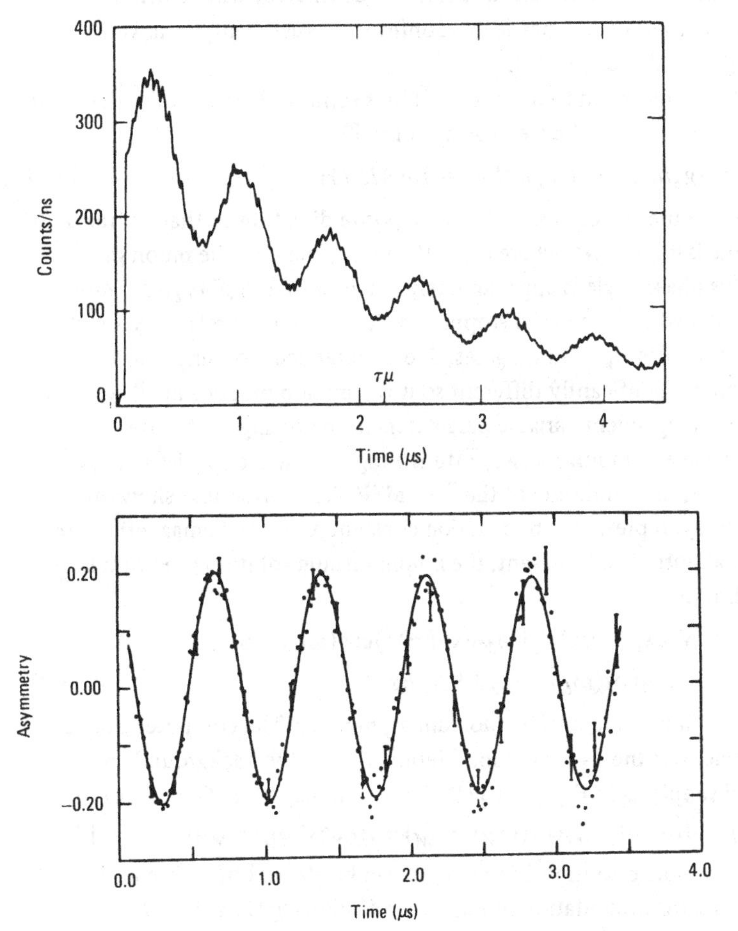

as ^{S}Mu and ^{T}Mu, respectively. It is expected that ^{S}Mu and ^{T}Mu will be formed initially in equal amounts as a result of the polarization of the muons and the random spins of the extracted electrons. Because of the hyperfine coupling and large magnetic moment of the electron, the muon spin in ^{S}Mu is rapidly depolarized by oscillation at the hyperfine frequency of 4463 MHz. This is too fast for current counting equipment, so ^{S}Mu is not observable by μSR. For free muonium atom studies, therefore, one can only measure ^{T}Mu and double its yield to include ^{S}Mu.

In the presence of transverse fields the triplet state of Mu will be resolved into three levels. Of the four transitions possible, two are at higher frequencies than practical while the two lowest (ν_{12} and ν_{23}) are essentially degenerate at fields <10 G (Brewer *et al.*, 1975). This is indicated qualitatively in a Breit-Rabi diagram such as Figure 3.5. It is quite convenient to study Mu at low fields and to have only one precession frequency. The degeneracy can be lifted at fields greater than 10 G when needed to confirm an assignment, as in water for instance (Percival *et al.*, 1976).

This low-field precession frequency of ^{T}Mu is some 103 times larger than that of μ^+, being dominated by the electron as in Eq. [3.8].

$$\omega_M = (g_e/m_e - g_\mu/m_\mu)eH/4c = 1394H \text{ kHz} \qquad [3.8]$$

Furthermore, the precession occurs in the opposite direction to that of free μ^+, so that θ_M equals $\phi_M - \omega_M t$, where ϕ_M is the initial phase of the muon in muonium. This phase angle is approximately equal to that in $\mu^+(\phi_D)$ for most condensed media where the muon stopping region subtends a relatively small solid angle at the telescope. But in gases, due to extended stopping volumes, ϕ_M and ϕ_D can be significantly different so it is common practice in all systems to use them as independent variable parameters in the fitting procedure.

Unlike diamagnetic muon species, ^{T}Mu is a highly reactive species in most media. Therefore, the amplitude of the ^{T}Mu MSR signal invariably shows an exponential decay, represented by reaction constant λ. Since diamagnetic states of μ^+ are also almost always present, the muonium spin rotation spectrum is of the form of Eq. [3.9],

$$N(t) = N_0 \exp(-t/\tau)[1 + A_M \exp(-\lambda t)\cos(\omega_M t - \phi_M) + A_D \cos(\omega_D t + \phi_D)] + N_0 Bg \qquad [3.9]$$

where A_M is the amplitude of the muonium asymmetry. The computer can be made to subtract out the natural muon lifetime factor, the background and to plot the signal simply as in Eq. [3.10] (Brewer & Crowe, 1978; Garner, 1979).

$$A(t) = A_M \exp(-\lambda t)\cos(\omega_M t - \phi_M) + A_D \cos(\omega_D t + \phi_D) \qquad [3.10]$$

A feeling for the time-evolution of this signal can be gleaned by reference to the geometric vector presentation in Figure 3.6 (following Garner, 1979).

In practice it is more accurate to obtain A_M and A_D from separate runs at different fields, because of the marked difference in the precession frequencies ω_M and ω_D. Figure 3.7(a) shows raw data obtained for water at 8 G, where the ^{T}Mu precession can readily be seen but that of D is too slow. This same sample run at 80 G (Figure 3.7(b)) shows the diamagnetic precession but ^{T}Mu is too fast to resolve (precession period ~9 ns at 80 G). Figure 3.7(c) shows the 8-G data fitted to Eq. [3.10].

Raw histograms are fitted to these equations by the χ^2-minimization program MINUIT (James & Roos, 1971) until the best fit is obtained. The best values of

Figure 3.5. A representative Breit-Rabi diagram for muonium, which is a two spin-1/2 system. The figure shows the effect of transverse fields on the energy levels resulting from the splitting of the triplet ($S = 1$). In low fields only the two transitions indicated by solid lines (ν_{12} and ν_{23}) are observable and these are degenerate at low enough fields. In the high-field limit the two transitions indicated by R are those observed for muonium-containing free radicals.

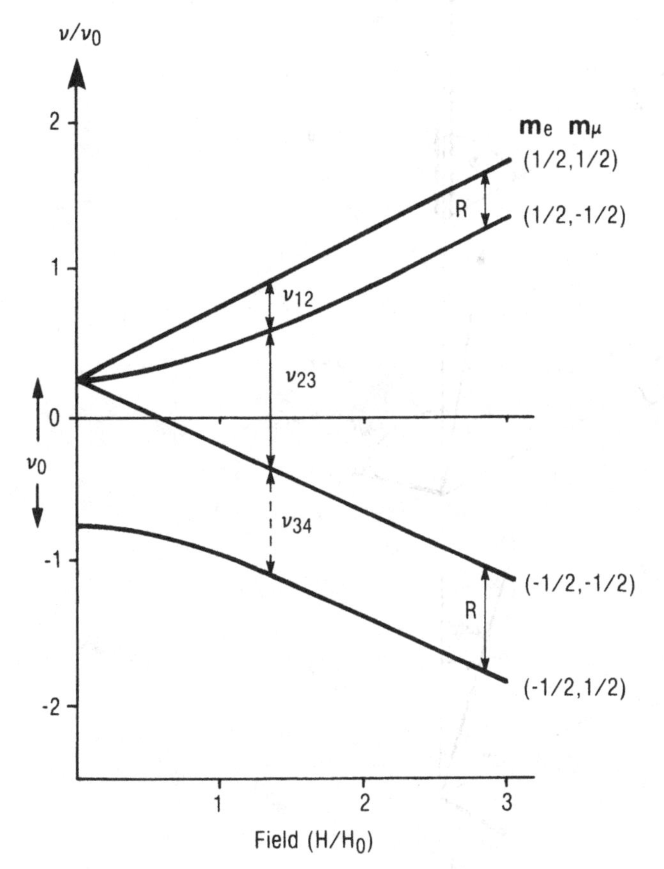

the parameters A_M, A_D and λ are those of interest in chemistry. The first two give measures of the yields of these two muon states in a given system, and λ represents the pseudo-first order rate constant by which ^TMu reacts.

3.4 Muonium-radical spin rotation (MRSR)

When Mu adds to a double bond, as in Eq. [3.11],

$$Mu + RCH{=}CH_2 \rightarrow R\dot{C}H{-}CH_2Mu \qquad [3.11]$$

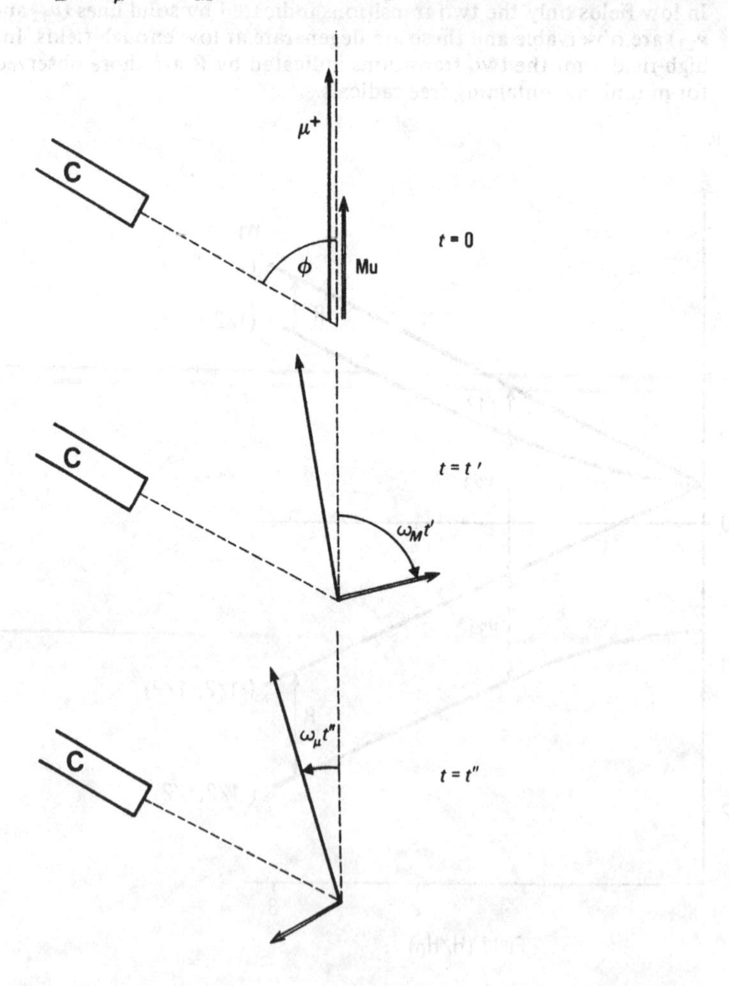

Figure 3.6. An attempt to indicate the time-dependence of the amplitude of the asymmetry as seen by the detector (C) at angle ϕ to the initial muon spin in diamagnetic states (μ^+) and in Mu. They precess in opposite directions with different Larmor frequencies $\omega_D \equiv \omega_\mu$ and ω_M, and the Mu amplitude falls due to chemical reaction.

a free radical is produced in which the muon is removed from the centre of unpaired electron density by at least one carbon atom. The muon-electron hyperfine coupling is, therefore, much weaker than in Mu; but the splitting in a magnetic field still follows the pattern of Figure 3.5. However, there is also

Figure 3.7. μSR lifetime histograms obtained in pure water at 295 K. (a) Raw data in 8-G transverse field; (b) raw data in 80-G field; (c) the data in (a) computer fitted to Eq. [3.10] with background and muon decay removed. Bearing in mind the Larmor frequencies of μ^+ in diamagnetic states and in muonium, one sees only the higher-frequency (11 MHz) muonium precession in (a) and only the lower frequency (~1.1 MHz) of diamagnetic μ^+ in (b). The amplitude of the oscillation in (b), A_D, is very much greater than the amplitude of the oscillation in (a), A_M, because some 60% of the muons in water form diamagnetic states while only 10% form ^TMu (see chapter 6).

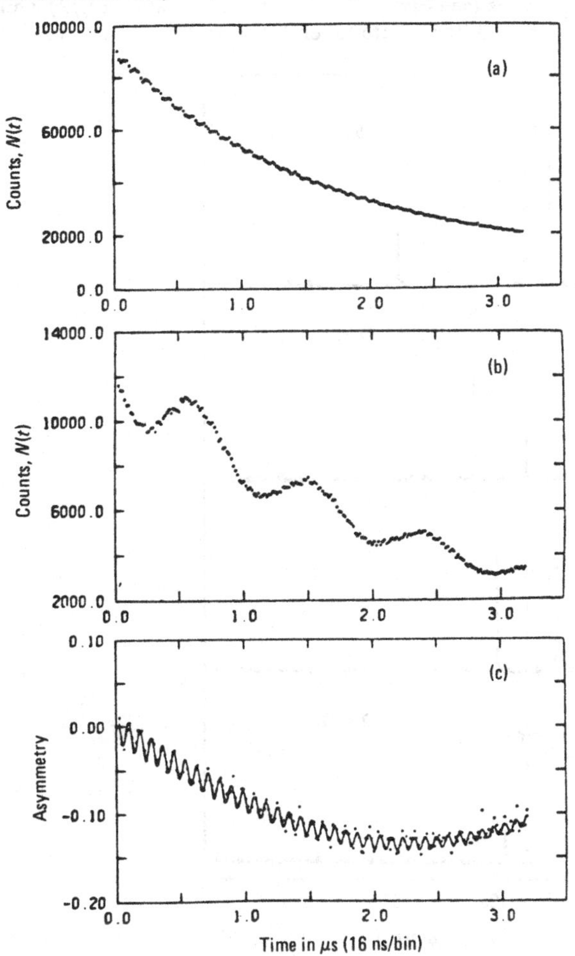

coupling with other magnetic nuclei, such as protons, when these are present in the radical. This causes the amplitude of the asymmetry to be spread out over many precession frequencies.

This difficulty was solved by Roduner *et al.* (1978) by decoupling the spins at high magnetic fields, where the electron Larmor frequency exceeds the muon and proton hyperfine frequencies. In the high field limit one gets just the pair of frequencies indicated by *R* in Figure 3.5 as in muonium, split symmetrically about the muon's Larmor precession frequency. Since there are invariably

Figure 3.8. Muon spin precession frequencies shown after Fourier transform, for 2,3-dimethyl-2-butene at four transverse magnetic fields (0.6–5 kG). *D* is the Larmor frequency of muons in diamagnetic environments (ω_D in text); *R* are the two muonium-containing free radical frequencies (see Figure 3.5) whose mean value ($\omega_r/2$) is indicated by the dotted line. (Data taken from Roduner *et al.*, 1978.)

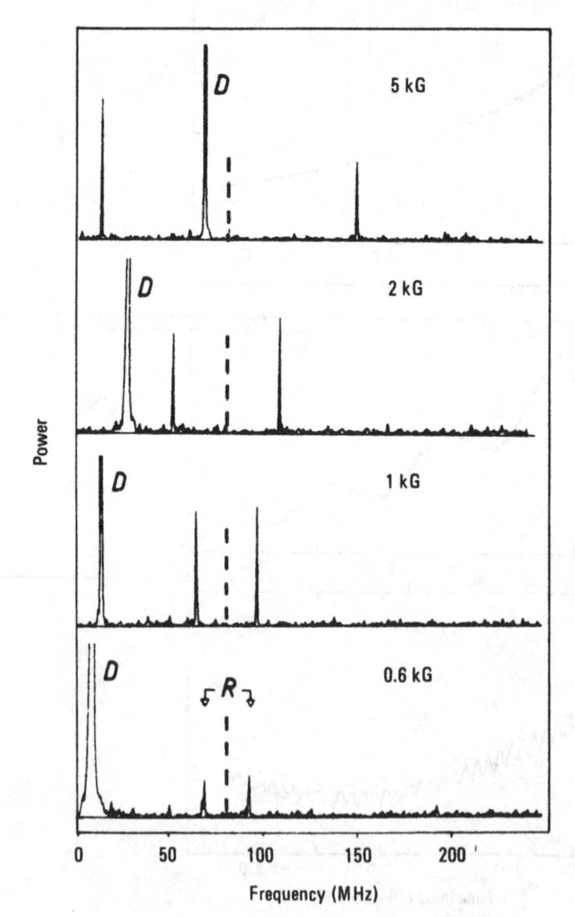

some diamagnetic muon states also formed in addition to the radical, one obtains a Fourier transform of the μSR spectrum as typified by that of Figure. 3.8.

The muon polarization can be fit to an equation of the form of [3.12]

$$N = N_0 \cos[(\omega_D + \omega_r/2)t] + \cos[(\omega_D - \omega_r/2)t] \quad [3.12]$$

because the two radical frequencies, whose sum is ω_r (taking due account of their different polarity) are displaced $\pm\,\omega_D$. Figure 3.8 shows the first Mu-radical spin rotation Fourier transform spectrum and demonstrates very clearly that the radical frequencies are centred at $\omega_r/2$ (dotted line) which is independent of field.

MRSR can be accomplished with apparatus as in Figure 3.3 just as in μSR and MSR only with much higher (0.5-5 kG) magnetic fields. One can measure both the amplitude (yields) and hyperfine coupling constants (ω_r or A_μ) of the radicals. The latter typically fall in the range 20 to 500 MHz, showing that the hyperfine splitting is much less than in free Mu. They happen to be higher than their corresponding H-atom analogues, however (see chapter 9).

3.5 Pulsed μSR

Until recently almost all μSR chemistry was conducted with a continuous flux of muons (d.c. mode) by observing the arrival of one muon at a time and timing its decay positron. In order to maintain individual muon-positron correlations a second muon cannot be accepted until the first has decayed. This necessarily limits the flux to about $10^5\,s^{-1}$ and therefore to $\gg 10^2$ s as the minimum time for a 10^7-event histogram.

However, many high-energy accelerators work in pulsed-mode, producing short intense bunches of particles - which in turn produce pulses of pions then muons - so μSR technology is beginning to take advantage of this. In Figure 3.9 are shown the two detection approaches adopted (Nagamine, 1981). These make use of 50-ns long pulses, with $\sim 10^4\,\mu^+$ per pulse, separated by 50 ms. Both digital and analogue detection methods are used. The digital method resembles the d.c.-mode in that, using many fast detectors, decay-positrons are individually timed and referred back to the centre of the pulse for the 'start' time. These lifetimes are binned to give a lifetime pattern in a multichannel digitizer. In the analogue method, a single large detector sums up all the positron pulses to give the overall shape of the positron lifetime distribution. A series of such shaped pulses are added together on a transient digitizer to produce a 10^7-event spectrum. In the latter method, photomultiplier saturation has to be avoided, in fact one is limited by the range of its dynamic linear response.

This pulsed mode has several advantages. First, the lifetime histogram can be observed out to much longer times (in fact to the pulse separation time) because of the very low background. Therefore one can examine particularly slow chemical reactions, spin-relaxations and μ^+ diffusions by this method. Second, it may soon become much speedier than the d.c. mode: indeed, in principle it is possible to obtain a μSR spectrum from a single pulse. Third, it allows one to pulse associated stimulation or detection equipment either concurrent with the muon pulse or at a predetermined time afterwards. This enables high-intensity pulsed r.f. generators to be used for muon spin resonance and opens the possibility of high-intensity laser stimulation or electric field studies. It may even be possible to study second-order chemical reactions of μ^+ or Mu (Mu_2 for instance). There are some disadvantages – discrimination against contaminants in the beam is particularly difficult, for example. But when very short (10^{-9} s), very intense, clean muon pulses become available at variable repetition rates, then the pulsed-mode will probably be the far superior method.

3.6 Longitudinal field studies – muon spin relaxation

A typical experimental arrangement for longitudinal (or zero) field studies is provided in Figure 3.10 and sketched in Figure 3.1(b). Muons enter the sample through holes in the backward counters (EB1 and EB2), stop in the sample, associate chemically, then start to decay (and possibly react). As they

Figure 3.9. Pulsed μSR apparatus showing the digital and analogue methods of analysis. (Figure taken from Nagamine, 1981.)

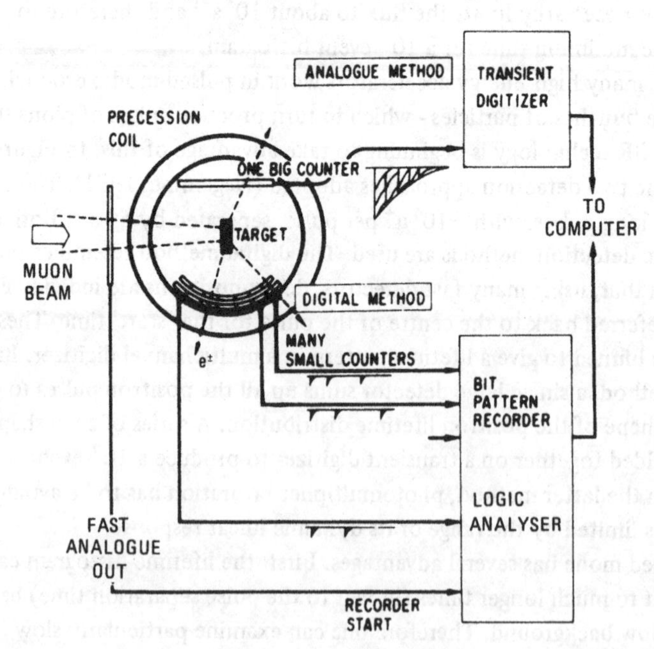

are spin polarized in the direction of the field there is no rotation: instead, their asymmetric decay gives a higher rate in the forward counters (EF1 and EF2) than in the backward counters when using 'backward' (positive helicity) muons. In the absence of any relaxation of the spin polarization, the forward and backward histograms merely show the muon decay and their ratio remains constant.

When only diamagnetic muon states are present, the asymmetry (A) and any spin-lattice relaxation or spin exchange interaction – collectively characterized by a relaxation time T_1 – may be evaluated by fitting to an equation of the form of [3.13].

$$N(t) = N_0 \exp(-t/\tau)[1 \pm A \exp(-t/T_1)] + N_0 Bg \qquad [3.13]$$

In practice it is often preferable to obtain the relaxation function by the difference-sum method (Hayano *et al.*, 1979) as in Eq. [3.14]

$$R = (N_f - \alpha N_b)/(N_f + \alpha N_b) = \bar{A} \exp(-t/T_1) \qquad [3.14]$$

where α is an empirical internal calibration depending on the actual geometry, and N_f and N_b are the forward and backward count rates in corresponding time bins. In zero field, the treatment is straightforward (Hayano *et al.*, 1979; Uemura, 1981) and has been used for a particularly wide time range of spin dynamic studies in solid state physics.

When Mu is formed there are additional depolarization considerations (Brewer *et al.*, 1975). In zero field, ^{T}Mu is stationary but ^{S}Mu oscillates between +1 and −1, averaging to 0. This results in the muon polarization being reduced

Figure 3.10. Apparatus for muon spin relaxation studies based on asymmetry measurements in a longitudinal field. (Taken from Yamazaki, 1979.)

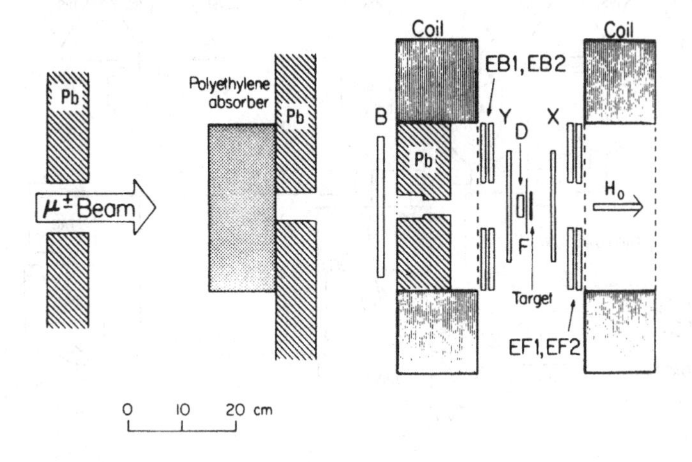

by $\frac{1}{2}h_M$, where h_M is the fraction of the muons initially forming Mu. In fields larger than the hyperfine field, this depolarization in ^{S}Mu is quenched by the decoupling of the hyperfine interaction, and one observes the fully polarization. At intermediate fields, ^{S}Mu oscillates between +1 and $(x^2-1)/(x^2+1)$ where $x = H/1585$ G. The observed muon polarization in longitudinal fields can thus be utilized through Eq. [3.15]

$$P_D(//^1) = h_D + h_M[1 + x^2/(x^2+1)]/2 \qquad [3.15]$$

where h_D is the fraction of muons initially forming diamagnetic states (Hughes, 1966; Percival, 1981).

Relatively few longitudinal-field chemical studies have been reported, though in combination with rotation they can be used to distinguish spin-exchange interactions from other types of chemical reactions. Indeed, Mobley *et al.* (1966, 1967) and Mobley (1967) originally used the field dependence of the polarization to demonstrate that reactions between Mu and O_2 or NO involved spin-exchange not chemical combination. The conversion of ^{T}Mu → ^{S}Mu shows up as a relaxation in longitudinal fields (because ^{S}Mu depolarizes its muon, except in

Figure 3.11. A simplified schematic representation of the MSR data acquisition logic used by Garner (1979). On the left is one positron telescope consisting of three counters (L1, L2 and L3) and on the right another (R1, R2 and R3). D is the muon 'start' counter. Each counter consists of a sheet of plastic scintillator coupled by a light pipe to a photomultiplier, whose power supply allows for energy discrimination (DISC).

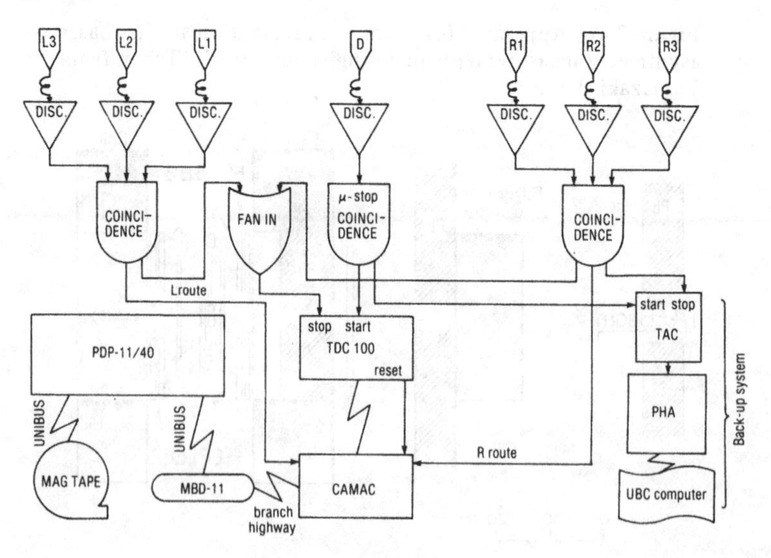

Figure 3.12. Layout of the Tri-University Meson Facility (TRIUMF) cyclotron and beamlines on the campus of the University of B.C., Vancouver, Canada. μSR chemistry is performed on the M20 channel. The cyclotron accelerates H^- to ~500 MeV, when the electrons are stripped off to produce bare protons which are automatically bent out of the cyclotron into a beamline. Beamline 1 (BL1) feeds the meson hall where two targets, T1 and T2, of Be, Cu or C produce intense fluxes of pions for collection in channels M13, M11, M9, M8 and M20.

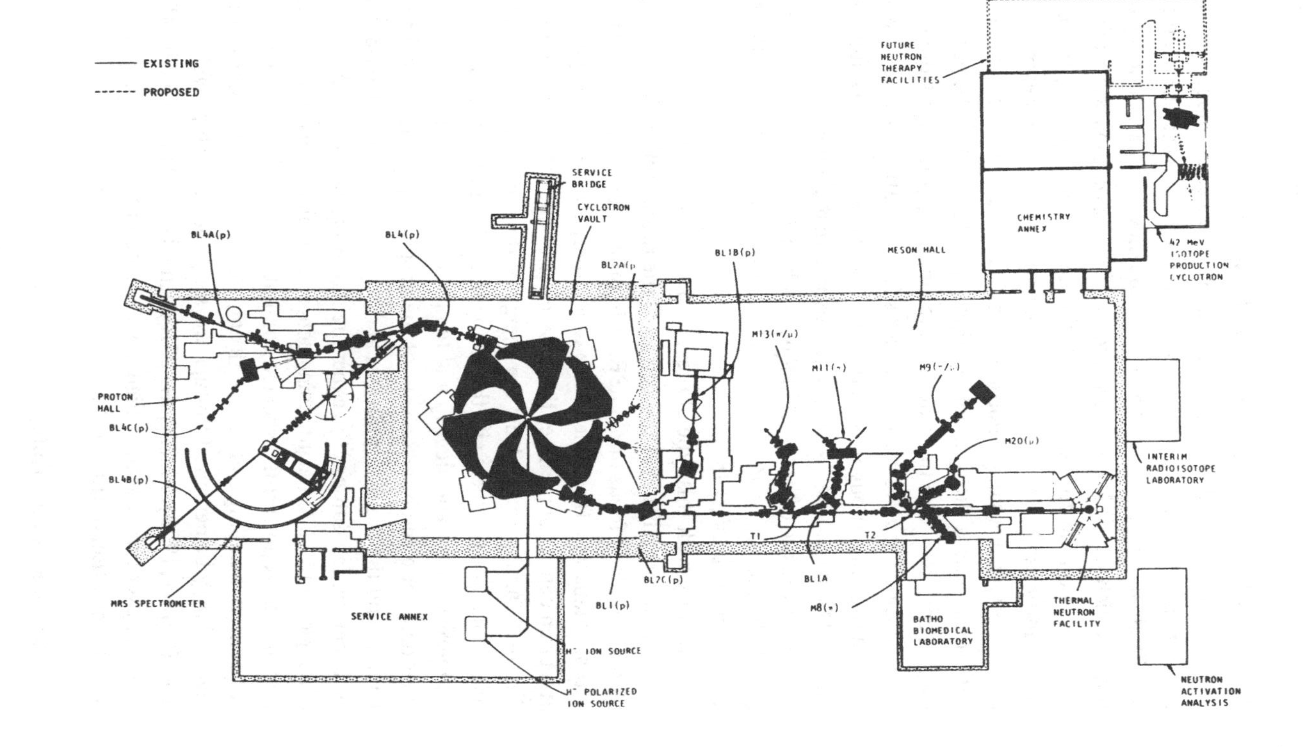

very high fields), whereas a conversion of $^{T}Mu \rightarrow \mu^{+}$ in a chemical reduction does not affect the longitudinal polarization (because the muon spin is unaffected).

3.7 Other techniques and practical matters

As indicated by Figure 3.1(c), one can also do muon spin resonance experiments. In this, one measures the forward and backward count rates in a longitudinal field as a function of the frequency of radiofrequency or microwave power which was fed to the target vessel. The forward-backward difference shows a maximum corresponding to the frequency of the hyperfine splitting. Its intensity can be related to the amount of ^{T}Mu. In the presence of scavengers its intensity changes in proportion to the rate constant for conversion $^{T}Mu \rightarrow {}^{S}Mu$ with relaxation of the spin-polarization. Therefore, it has the potential of being used in chemistry like the other μSR techniques; yet, apart from the early work of Ziock *et al.* (1962) and Hughes (1966), rather little chemistry has been studied this way, though it is certainly being exploited successfully in solid state physics (Yamazaki, 1979). It is also possible that chemists may soon show interest in the use of other techniques such as double electron muon resonance (DEMUR) (Brown *et al.*, 1979; Vanderwater *et al.*, 1981) or μ^{-}SR (Nagamine, 1979; Suzuki *et al.*, 1980).

Difficulty with this type of presentation is encountered because one has to survey and generalize, yet this can leave the reader with no concrete experimental details to focus on. In μSR, as in other sophisticated techniques, the actual details vary from one centre to another, even from one year to the next at a given centre. Sometimes this is by choice; but often it is imposed by different parameters, such as: the energy of the muon beam; the duty cycle of the machine producing it; the contamination in the beam; its size and luminosity; the shape, density and temperature of the sample being studied; the orientation and homogeneity of the magnetic field being applied to the sample; and the number, positions and types of positron telescopes being used as detectors. In addition, the experimental set-up requires elaborate electronics and logic, with pulse height, coincidence, anticoincidence and veto discrimination modes. A powerful dedicated computer is virtually essential for data acquisition, handling and on-line fitting. Some idea of the ramifications involved may be perceived through the simplified sketch provided as Figure 3.11 (Garner, 1979). Finally, the μSR channel (M20) at the TRIUMF cyclotron in Vancouver can be seen in Figure 3.12 to be but a small component of the overall operation of that facility.

References for chapter 3

Brewer, J. H., Crowe, K. M., Gygax, F. N. & Schenck, A. (1975). Positive muons and muonium in matter. In *Muon Physics*, vol. 3, ed. V. W. Hughes & C. S. Wu, pp. 3–139. New York: Academic Press.

Brewer, J. H. & Crowe, K. M. (1978). Advances in muon spin rotation. *Annual Review of Nuclear and Particle Science*, **28**, 239–326.

Brown, J. A., Heffner, R. H., Leon, M., Dodds, S. A., Vanderwater, D. A. & Estle, T. L. (1979). Detection of EPR transitions of muonium in quartz by muon-spin rotation. *Physical Review Letters*, **43**, 1751–4.

Garner, D. M. (1979). Application of the muonium spin rotation technique to a study of the gas phase chemical kinetics of muonium reactions with the halogens and hydrogen halides. Ph.D. thesis, University of British Columbia, *Dissertation Abstracts*, 1–307.

Garwin, R. L., Lederman, L. M. & Weinrich, M. (1957). Observations of the failure of conservation of parity and charge conjugation in meson decays: the magnetic moment of the free muon. *Physical Review*, **105**, 1415–17.

Hayano, R. S., Uemura, Y. J., Imazato, J., Nishida, N., Nagamine, K., Yamazaki, T. & Yasuoka, H. (1978). Longitudinal spin relaxation of μ^+ in paramagnetic MnO. *Physical Review Letters*, **41**, 421–4.

Hayano, R. S., Uemura, Y. J., Imazato, J., Nishida, N., Yamazaki, T. & Yasuoka, H. (1979). Zero- and low-field μ^+ spin relaxation behaviour in MnSi. *Hyperfine Interactions*, **6**, 133–6.

Hughes, V. W. (1966). Muonium. *Annual Review of Nuclear Science*, **16**, 445–70.

James, F. and Roos, M. (1971). MINUIT, CERN Computer. 7600 Interim Program Library.

Mobley, R. M., Bailey, J. M., Cleland, W. E., Hughes, V. W. & Rothberg, J. E. (1966). Muonium chemistry. *Journal of Chemical Physics*, **44**, 4354–5.

Mobley, R. M., Amato, J. J., Hughes, V. W., Rothberg, J. E. & Thompson, P. A. (1967). Muonium chemistry II. *Journal of Chemical Physics*, **47**, 3074–5.

Mobley, R. M. (1967). Interactions of muonium with atoms and molecules. Ph.D. thesis, Yale University, *Dissertation Abstract:* 67-8399, pp. 1–92.

Nagamine, K. (1979). Negative muon spin rotation in solids. *Hyperfine Interactions*, **6**, 347–55.

Nagamine, K. (1981). Pulsed μSR facility at the KEK booster. *Hyperfine Interactions*, **8**, 787–96.

Ng, B. W. (1980). *Studies of the Muonium Atom in Liquid Media*. M.Sc. Thesis, University of British Columbia.

Percival, P. W. (1981). The missing fraction in water. *Hyperfine Interactions*, **8**, 325–8.

Percival, P. W., Fischer, H., Camani, M., Gygax, F. N., Ruegg, W., Schenck, A., Schilling, H. & Graf, H. (1976). The detection of muonium in water. *Chemical Physics Letters*, **39**, 333–5.

Pifer, A. E., Bowen, T. & Kendal, K. R. (1976). A high stopping density μ^+ beam. *Nuclear Instrumental Methods*, **135**, 39–46.

Roduner, E., Percival, P. W., Fleming, D. G., Hochmann, J. & Fischer, H. (1978). Muonium-substituted transient radicals observed by muon spin rotation. *Chemical Physics Letters*, **57**, 37–40.

Suzuki, T., Mikula, R. J., Garner, D. M., Fleming, D. G. & Measday, D. F. (1980). Muon capture in oxides using the lifetime method. *Physics Letters*, **95B**, 202–7.

Swanson, R. A. (1958). Depolarization of positive muons in condensed matter. *Physical Review*, **112**, 580–6.

Uemura, Y. J. (1981). Probing spin glasses with zero-field μSR. *Hyperfine Interactions*, **8**, 739–48.

Vanderwater, D. A., Dodds, S. A., Estle, T. L., Brown, J. A., Heffner, R. H., Leon, M. & Cooke, D. W. (1981). DEMUR: Double electron muon resonance. *Hyperfine Interactions*, **8**, 823–6.

Yamazaki, T. (1979). Muon spin relaxation in magnetic materials. *Hyperfine Interactions*, **6**, 115–25.

Ziock, K., Hughes, V. W., Prepost, R., Bailey, J. M. & Cleland, W. E. (1962). Hyperfine structure of muonium. *Physical Review Letters*, **8**, 103–5.

4

ANALYSIS AND INTERPRETATION OF μSR DATA

The positive muon is observable through its magnetic moment and the interactions of its nuclear spin with any unpaired electrons in its immediate environment, as described in chapter 3. The purpose of the present chapter is to describe the use and scope of these data for studies in chemistry. There are three distinct types of chemical associations of μ^+ which are readily detectable by μSR: diamagnetic states (for convenience collectively given the subscript, D), free muonium atoms (Mu or subscript, M), and Mu-substituted free radicals (subscript, R). Each has its characteristic, field-dependent, precession frequency, and in practice these are so far apart that the three states can be studied separately at different fields.

All μSR histograms are accumulated from one-at-a-time events (except in pulsed mode), so that a vast number of isolated events are summed together and it is assumed this is equivalent to the time distribution that a large instantaneous ensemble would have. There is always the overall exponential fall-off ($\exp(-t/\tau)$) in the number of muons remaining with the progress of time, because of the natural decay of the muons by which one observes their spins. But this decay is of no interest, so the computer is made to remove it at an early stage in the analysis of data. The decay that is mainly of interest in chemistry is represented by λ in Eq. [3.10]. This refers to muonium; but, in general, any of the observable species may be seen to decay, separately or together, by the use of an overall fitting equation such as [4.1],

$$N(t) = N_0\{Bg + \exp(-t/\tau)[1 + \Sigma A_i \exp(-\lambda_i t)\cos(\omega_i t + \phi_i)]\} \quad [4.1]$$

where A_i, λ_i, ω_i and ϕ_i are the amplitudes (asymmetries), decay constants, precession frequencies and initial phases, respectively, of the different muon states.

The histograms of Figures 3.4 and 3.7 cover the μSR observation time-window from 10^{-7} s to several microseconds. However, it is important to note that in order to register in the coherent oscillations from 10^{-7} to 10^{-5} s, the species must have been *in phase* initially (same ϕ_i) – or, at least, the muon must not have undergone prior precession at a different frequency. Signals which are out-of-phase due to interconversions between 10^{-9} and 10^{-7} s merely contribute to the random background count rate, *Bg*. This is shown in Figure 4.1(a).

4.1 Diamagnetic muon states (D)

This component of Eq. [4.1] is simplified by the fact that free μ^+, or diamagnetic molecules containing it, are not normally reactive on the timescale of μSR observations (10^{-7}–10^{-5} s). Muons seem to either react or associate chemically in a much shorter time than 10^{-7} s, or else they remain in diamagnetic states for times well beyond 10^{-5} s. There are a few exceptions, such as in pure CS_2 (Ng, 1980) or strongly paramagnetic solutions (Schenck *et al.*,

Figure 4.1(a). The dephasing effect of a chemical reaction Mu → D occurring after precession has set in.

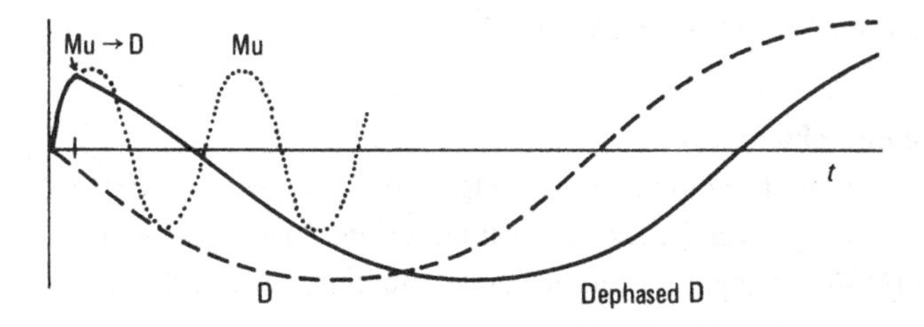

Figure 4.1(b). Schematic representation of the information contained in an MSR histogram. Of prime interest are λ_M, ω_M, and A_M. These data refer to 10^{-4} M phenol in water. The exponential muon decay and the background *Bg* have already been removed.

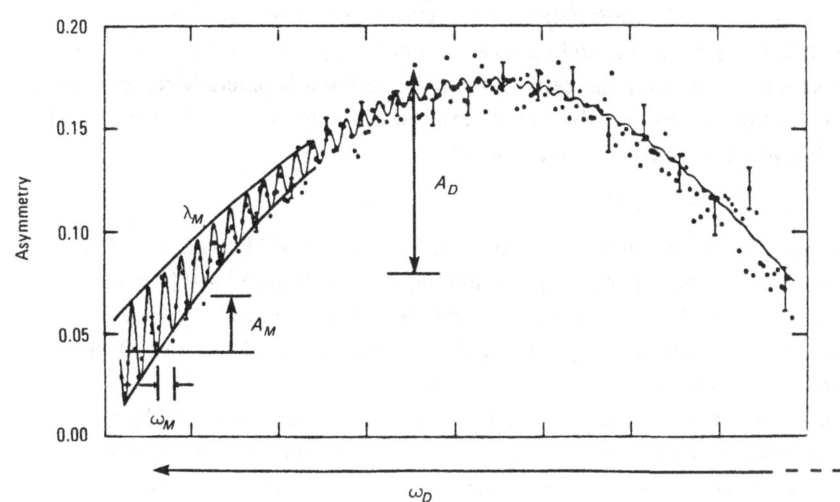

1972), otherwise diamagnetic states convert only to other diamagnetic states - all of which have the same precession frequency (within present resolution times). In general, then, $\exp(-\lambda_D t) \rightarrow 1$ and the diamagnetic μSR signal fits Eq. [4.2] after removal of Bg and the muon decay.

$$A(t) = A_D \cos(\omega_D t + \phi_D) \qquad [4.2]$$

A_D is the time-zero amplitude of the signal arising from coherently precessing muons in all diamagnetic states, including: free, trapped or solvated μ^+, and diamagnetic molecules such as MuH. (A_D is employed here instead of A_μ as formerly used. There are three reasons for changing. (i) A_μ is currently being used for the Mu-radical hyperfine coupling constant by analogy with the proton counterpart in ESR, A_p (Roduner *et al.*, 1978; Roduner & Fischer, 1981). (ii) Although no distinction is made at present between free μ^+ and diamagnetic molecules, that may be possible in the near future, and the asymmetry of each will need to be specified separately. (iii) The subscript D is already in use with the fractional asymmetries, P_D (see later).)

4.2 Free muonium atoms

In muonium, the muon spin is coupled not only to the external field but to the electron spin via the hyperfine interaction. As a result, a muonium spin rotation (MSR) histogram is, in principle, much more complicated than is μSR for the diamagnetic cases above.

Furthermore, Mu is a highly reactive chemical species. It attacks some solvents, but fortunately is stable in several saturated inert liquids like water and alkanes. To these can be added solutes at concentrations where the Mu reactions occur mainly in the 10^{-7}–10^{-5} s observation-window. A typical MSR histogram has the features indicated in Figure 4.1(b). The damping of the oscillation is exponential and yields a coefficient λ_M. This is occasionally just a relaxation of the spin-polarization; but in chemistry it is generally representative of a chemical reaction of Mu. In the presence of solute, S, at concentration $[S]$ the Eq. [4.3] is generally found to apply,

$$\lambda_M = \lambda_0 + k_M[S] \qquad [4.3]$$

where k_M is the bimolecular rate constant for reaction of Mu with S, and λ_0 is the 'background' value of λ_M as found in the solvent at $[S] = 0$ (see later). Since the method utilizes one muon at a time in the cell (except for pulsed μSR), these experiments occur under the ideal conditions of 'infinite dilution' with $k_M[S]$ a pseudo-first order rate constant.

Many muonium studies focus just on the kinetic data represented by k_M. These describe the 'chemical reactivity' of Mu, and can be related to the corresponding k values of the heavier isotope, H. Such k_M values are conveniently and unambiguously obtained from the empirical fits of the actual μSR histograms. Therefore they arise from direct observations, on the 10^{-7}–10^{-5} timescale.

The actual evolution of the time-dependence of the muon polarization in muonium has been treated fully in several articles during the past decade (Gurevich *et al.*, 1971; Brewer, Crowe, Gygax & Schenck, 1975; Schenck, 1976; Percival & Fischer, 1976; Fleming *et al.*, 1979; Garner, 1979). In ^{S}Mu the muon's spin oscillates at the hyperfine frequency (4463 MHz) so that the total polarization from ^{S}Mu + ^{T}Mu follows the trace in Figure 4.2. Current detectors are unable to follow the fast oscillation: instead, they merely display the mean value given by the dotted line in Figure 4.2. This is the 1.39 MHz G^{-1} frequency (Eq. [3.8]) seen in an MSR histogram such as Figure 3.7(c). It gives a measure of ^{T}Mu, with ^{S}Mu averaged to zero.

At sufficiently low magnetic fields (<10 G) the normal MSR signal fits quite well to Eq. [4.4].

$$A(t) = A_M \exp(-\lambda_M t)\cos(\omega_M t - \phi_M) \qquad [4.4]$$

There is an additional beat frequency term but this is normally negligible on fits out to only 4 μs (Garner, 1979). At intermediate fields (10–150 G) the low-field degeneracy of ν_{12} and ν_{23} (Figure 3.5) is lifted and a two-frequency precession of

Figure 4.2. The time-evolution of the muon spin polarization in muonium in a 100-G transverse field. In practice, current detectors cannot follow the high frequency and respond in accordance with the average value shown by the dotted line. (Figure taken from Garner, 1979.)

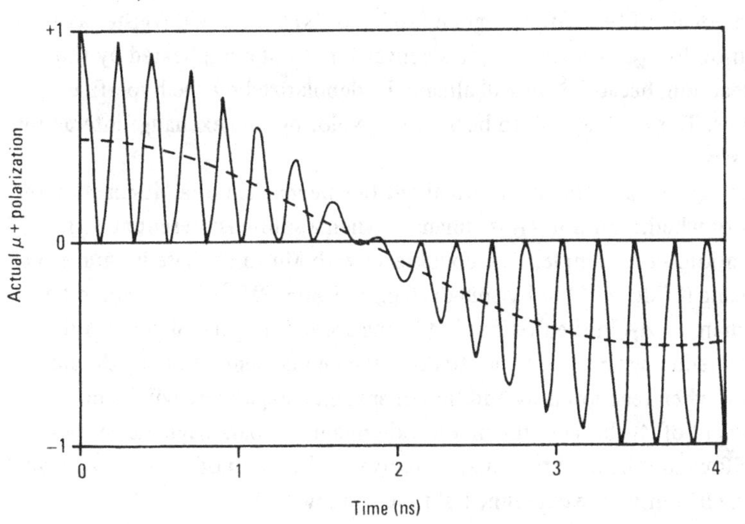

Mu pertains, as originally observed in quartz (Gurevich *et al.*, 1971). For fields greater than ~150 G the Mu precession frequency exceeds the current experimental resolution.

4.3 Interconversion of Mu and μ^+ (or D)

The requirement of coherent precession by all muon spins contributing to A_D or A_M, as mentioned in the preamble of this chapter, has several important uses. Most of them are unique aspects of μSR as a technique. They are listed briefly here, though some of them will be used and developed more cogently later. In practice the most common form of conversion is from Mu → μ^+ with possible enhancement in A_D induced by the presence of a solute. But the reverse process, with enhancement in A_M, should also be considered. (It is, presumably, responsible for the presence of Mu in the first place.)

(a) If the product of a conversion Mu → μ^+ (or D) is to contribute to the observed A_D, then the reaction time has to be much shorter than the Mu precession period in the applied field. Thus, if A_D is being measured in a 100-G field then the Mu precession period is some 7 ns (see Eq. [3.8]), so the reaction time should be much less. Even for a solute with k_M close to the diffusion-limited value of $\sim 10^{10}\,\text{M}^{-1}\,\text{s}^{-1}$, this means $[S]$ must be 0.1 M or higher. Since this applies only to one half of the Mu ensemble (^{T}Mu), A_D could increase by a maximum of $A_M/2$.

(b) If conversion of the other half, ^{S}Mu → μ^+, is to be included, then the reaction time must be much less than the hyperfine period, 0.2 ns. This normally requires solute concentrations > 1 M.

(c) The conversion ^{T}Mu → ^{S}Mu can result from an electron spin-exchange interaction, or spin-flip. If this occurs in the presence of a paramagnetic solute on the timescale of the μSR histogram (10^{-7}–10^{-5} s) then it will register as a contribution to λ_M. Furthermore, this conversion is not complicated by the reverse reaction, because ^{S}Mu will already be depolarized by the hyperfine oscillations. This enables μSR to be used for exploring spin-exchange interactions in chemistry.

(d) Inferences can often be drawn about the mechanism of a Mu-reaction on the basis of whether or not A_D is enhanced when a solute is present at high concentration. For example, Fe^{3+} could react with Mu to dephase its spin either by oxidizing it (Mu → μ^+) or by spin-exchange (^{T}Mu → ^{S}Mu). There would be enhancement of A_D in the first but not in the second. In general, when Mu reacts as a reducing agent or by abstraction then enhancement of A_D should occur; but when reaction is by addition or spin-exchange there will be no enhancement of A_D because the muon will remain in a paramagnetic species.

(e) μSR cannot be used to actually observe the build-up of a reaction product containing μ^+ during the experimental time-window (10^{-7}–10^{-5} s). This is because the products of reaction will be formed at statistically distributed reaction times, and therefore the initial muon spins in the products will have been randomized on the very timescale of observation. If the muons were in the same magnetic states (presumably diamagnetic ones) in both reactants and products, then no detectable change in the μSR histogram would arise from the reaction.

(f) For conversions from $\mu^+ \rightarrow$ Mu the timescale limitations are much less stringent than in (a) above. This is because the measurement of A_M is conducted at lower fields than for A_D, so the precession periods are longer. Dephasing in this case arises from the variations in the precession of the product (but, again, Mu) as a result of the distribution of reaction times.

(g) The actual magnitude of the enhancement in A_D arising from conversion (Mu $\rightarrow \mu^+$) can be calculated from the known timescales; or, alternatively, the measured enhancement can be used to calculate unknown reaction rates. Both of these have been exploited quantitatively. Until a few years ago, evidence for the existence of Mu in liquids was based solely on inferences drawn from the measured enhancement in A_D caused by the presence of additives (Firsov & Byakov, 1965; Ivanter & Smilga, 1968; Brewer, 1972; Brewer *et al.*, 1971, 1973, 1974). It became possible to deduce reaction mechanisms and postulate Mu-radical intermediates, based on the extent to which the total muon spin polarization was observed to be conserved in concentrated solutions, and on

Figure 4.3. Typical 'titration' curves used to deduce the existence of Mu and its reaction mechanism for solutions of I_2 in methanol at 100 G. The top plot shows P_D and the bottom curve $\Delta\phi_D$ as a function of $[I_2]$. (Figure taken from Brewer *et al.*, 1974.)

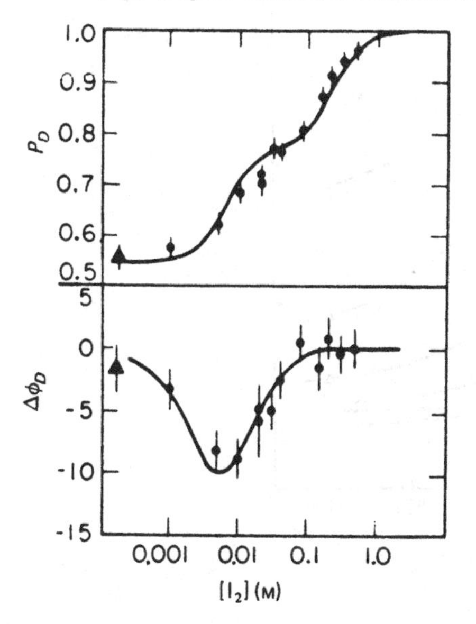

the variation of initial phase (Brewer *et al.*, 1974). An example of the type of 'titration' curve obtained in these early studies on Mu in liquids is given as Figure 4.3. This model-dependent, indirect, method has been superseded by the direct observation of Mu. But another exploitation of the quantitative enhancement of A_D for the deduction of mechanism is based on its magnetic field dependence at various high concentrations of Mu scavengers. Their general form is illustrated by the plots of Figure 4.4 (Percival, 1981).

In the presence of a solute converting Mu → D, the value of A_D is increased above the pure solvent value in accordance with Eq. [4.5].

$$A_D(s) = A_M \int_0^t \exp(-\lambda_M t')\cos(\omega_M t' + \omega_D t')\,dt' \qquad [4.5]$$

Upon integration, inclusion of the imaginary part, and noting that $\omega_0 \gg \lambda_M \sim \omega_M \gg \omega_D$ leads to Eq. [4.6]

$$A_D(s) = A_M \cos(\omega_D t - \phi_D)(\lambda_M^4 + \lambda_M^2\omega_M^2)^{1/2}/(\lambda_M^2 + \omega_M^2) \qquad [4.6]$$

Figure 4.4. Variation of P_D with magnetic field for solutions of a solute ($S_2O_3^{2-}$) which converts Mu → D in water, as a function of magnetic field (H) at various concentrations. (From top to bottom 1.0, 0.4, 0.2, 0.08, 0.04 and 0.01 M.) (Figure taken from Percival, 1981.)

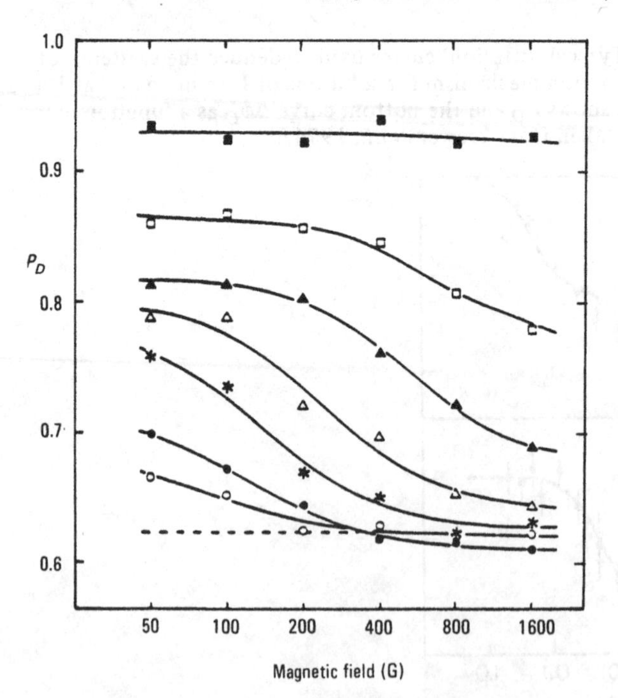

Now one can see that the whole signal $(A_D + A_D(s))$ arises from precessions at the diamagnetic frequency; and when $\lambda_M \gg (\omega_M + \omega_D)$ the total asymmetry equals $A_D + A_M$. Furthermore, in the limit when λ_M also greatly exceeds the hyperfine frequency, then ^{S}Mu is converted to μ^+ (or D) as well as ^{T}Mu, and the total diamagnetic asymmetry equals $A_D + 2A_M$. Providing there is no radical formation or depolarization occurring at times shorter than $1/\lambda_M$, this means that the total muon polarization appears in the diamagnetic fraction. This occurs with strong oxidizing agents at >1 M concentration in water, for instance.

4.4 Muonium-containing free radical studies

The direct observation of Mu-radicals by μSR as developed by Roduner *et al.* (1978) and Roduner & Fischer (1981) has rendered obsolete the earlier attempts to infer their presence by model-fitting procedures based on the enhancement of A_D as mentioned in (3) above. The direct method (MRSR) allows several aspects of Mu-radicals to be studied: their structure, isotope effects, kinetics, yields and formation mechanism. Each of these aspects will be outlined here.

First, on the matter of structure, this can often be deduced from the measured hyperfine coupling constants by analogy with the H-atom counterpart obtained by ESR. In the high-field limit, the sum of the two muon precession frequencies equals the hyperfine coupling constant A_μ as in Eq. [4.7],

$$|\nu_{12}| + |\nu_{43}| = A_\mu \ (= \omega_r/2\pi) \qquad [4.7]$$

bearing in mind that ν_{12} and ν_{43} (Figure 3.5) are predicted to be of opposite sense. This applies where the electron Larmor frequency greatly exceeds the hyperfine frequency when one has just the two frequencies and the energies are changing nearly linearly with field (*R* in Figure 3.5). In moderate fields (0.3 to 1 kG) other magnetic nuclei in the radicals, such as protons, split these muon frequencies to a measurable degree and this sometimes enables one to make further deductions about the structure of the radical. In weak fields the spin Hamiltonian includes important terms for each nuclear moment, so the radical's μSR spectrum is unduly spread out over many frequencies due to secondary splittings. But at zero field the situation is again quite tractable, particularly when there are only a few magnetically equivalent nuclei other than the muon. Such studies have been done with the earth's magnetic field compensated to less than 0.2 G (Roduner, 1979).

Second, important isotope effects have been found when A_μ is compared with the corresponding proton radical's coupling constant, A_p. The ratio A_μ/A_p is invariably greater than the ratio of muon to proton magnetic moments (3.18, see Table 3.1). This suggests secondary dynamic isotope effects play an important role. Such effects have been predicted to arise from the higher zero-point vibrational energy of the lighter isotope (Fessenden, 1964), and its smaller moment of inertia (Roduner, 1979).

Third, several types of kinetic effects of Mu-radicals have been studied. (i) In a molecule containing more than one double bond, such as styrene, only one Mu-radical ($C_6H_5\dot{C}HCH_2Mu$) is found to be present on the μSR timescale ($\sim 10^{-6}$ s) (Stadlbauer *et al.*, 1981). This raises the questions: is that the only radical found initially? or do the others convert to it by intramolecular relaxations? or did the others decompose in less than 10^{-7} s? It is not clear whether the selectivity arises from kinetic or thermodynamic considerations. (ii) In a similar vein one can study mixtures of unsaturated compounds, such as styrene and benzene, and examine the competition for the radical precursor. But, again, the relatively long period between formation ($<10^{-9}$ s) and observation ($>10^{-7}$ s) allows for the possibility that thermodynamic factors may alter the original kinetic preference. (iii) The actual reaction kinetics of the Mu-radicals towards additives can also be studied – either by observing the decay of the muon polarization at the radical frequencies, or by evaluating the radical lifetime from the linewidths (Roduner, 1981).

Fourth, the yields of Mu-radicals can be determined simply from the absolute measure of the total muon asymmetry at the radical frequencies. The yields, as one would expect, change with the properties of the medium. Any decomposition or rearrangement between 10^{-9} and 10^{-7} s will cause dephasing of the muon spins and result in lost polarization.

Fifth, is the matter of how the Mu-radicals are formed. Basically there seem to be three options. The radicals may be formed by addition of thermalized Mu atoms in ordinary thermal reactions before significant dephasing has occurred; or the radicals could be formed directly by hot (epithermal) Mu attack during the muon's thermalization process; or they could result from the mélange of ions, electrons, free radicals and excited states present at the very end of the muon's track.

4.5 Yields of muon states

The asymmetry (amplitude) parameters A_i which appear in the semi-empirical fitting equations (such as [4.1]) depend on several experimental variables, as noted in chapter 3, including: the polarization of the muon beam; the size and solid angle subtended by the counter telescopes; the amount of positron degrader used to enhance the effective positron anisotropy; plus geometrical factors associated with the shape, thickness, density and positioning of the sample. In order to eliminate these individual variations, it is necessary to have an internal calibration so that fractional, or normalized, yields can be quoted rather than measured asymmetries. Since liquid CCl_4 and metallic Al, for instance, always seem to show all incident muons as diamagnetic states, it has become common practice to use these as reference materials and to equate the value of A_D measured with CCl_4 (for liquid samples) and Al (for solids) with A_{max} for the particular experimental set-up used.

Fractional yields (P) for diamagnetic, free muonium and Mu-radical states are then calculated as follows: $P_D = A_D/A_{max}$; $P_M = 2A_M/A_{max}$; and $P_R = A_R/A_{max}$ where the 2 in P_M takes account of the fact that the ^{S}Mu half of the Mu ensemble is not observable, and A_R is the sum of the asymmetry distributed between the radical frequencies. In the case of radicals there may be several different contributing radical yields. In all liquids studied so far except CCl_4, and in some gases and solids, the value of $(P_D + P_M + P_R)$ is less than unity. This means that some of the original muon polarization is not observable at any of the frequencies studied due to unknown spin-depolarization or dephasing processes prior to 10^{-7}s; so there is a 'lost' or 'missing' fraction, P_L, given by Eq. [4.8].

$$P_L = 1 - P_D - P_M - P_R \qquad [4.8]$$

4.6 Transverse versus longitudinal field studies

Muon spin rotation studies in transverse magnetic fields are particularly versatile: one identifies the magnetic state, measures its yield, and determines its decay rate, all from the characteristic precession signals. In kinetic studies, it is the loss of phase coherence in the precession of the muon spins in ^{T}Mu as they react chemically that one monitors through λ_M. But all reactions which convert ^{T}Mu into something else contribute equivalently in transverse field measurements. In longitudinal or zero fields, on the other hand, the muon polarization – and its change with time – is determined by comparing the 'forward' and 'backward' positron counter-rates. The muon's spin in ^{S}Mu oscillates at the hyperfine frequency to give an equal positron rate in the forward and backward directions; but in ^{T}Mu it remains in the forward direction. This population difference decays with time for processes or chemical reactions causing a relaxation of the original spin polarization in ^{T}Mu.

In muonium chemistry it is often necessary to distinguish a chemical reaction, such as addition or reduction, from a spin-exchange interaction. This is important when comparing Mu with H rate constants, because the latter are not measured in a manner that includes spin-exchange. But if one merely used transverse field μSR data the value of k_M would include spin-exchange. Consider the reactions of ^{T}Mu with NO. There can be a combination reaction (Eq. [4.9]) or spin-exchange (Eq. [4.10]).

$$^T\text{Mu} + \text{NO} \xrightarrow{k_9} \text{MuNO} \qquad [4.9]$$

$$^T\text{Mu} + \text{NO}(\uparrow) \xrightarrow{k_{10}} {}^S\text{Mu} + \text{NO}(\downarrow) \qquad [4.10]$$

Both reactions contribute to λ_M from μSR in transverse fields: reaction [4.9] because the precession frequency of the muon changes on reaction so that it becomes dephased, and [4.10] because the muon is depolarized by the hyperfine oscillations in ^{S}Mu. Therefore the sum of these rate constants is measured: $k_9 + k_{10} = (\lambda_M - \lambda_0)/[NO]$, providing the diffusion-controlled limit is not reached by summing these *k*s. If this reaction between Mu and NO is now also studied in longitudinal field then these two *k* values can be separated. Reaction [4.9] does not affect the muon spin polarization, it merely breaks the hyperfine coupling and places the muon in a different magnetic state. Therefore, [4.9] does not cause depolarization, and k_9 does not contribute to the T_1^{-1} relaxation coefficient obtained from longitudinal field studies. The occurrence of reaction [4.10], however, depolarizes the muon spin by forming ^{S}Mu and therefore k_{10} alone is equated with T_1^{-1}. It is this comparison of the rate data obtained in transverse and longitudinal field modes that is particularly valuable in deducing reaction mechanisms.

4.7 'Background' relaxation or decay, λ_0, of muonium signals

λ_0 of Eq. [4.3] corresponds to a 'background' decay rate of the muonium precession signals. It is troublesome because it limits the accuracy with which small values of λ can be measured. Until very recently, λ_0 was always found to be $(2.5 \pm 0.4) \times 10^5 s^{-1}$, corresponding to a muonium lifetime of 4 μs. This same value was found in materials as diverse as water (Percival *et al.*, 1977; Jean *et al.*, 1978), liquid neopentane (Ng *et al.*, 1982; Miyake *et al.*, 1982), and N_2 gas (Garner, 1979). In water, this same value has been obtained under a variety of conditions: by different water purification technique; different experimenters in different laboratories, using a variety of magnets with different homogeneities; over a range of magnetic fields though mainly at 7-9 G. Furthermore, λ_0 shows an extremely small temperature dependence: $E_a < 1$ kJ mol^{-1}, which is barely beyond the error limits (Ng *et al.*, 1981).

Several possible causes have been considered. (i) The background decay seems unlikely to have arisen from impurities because the same value was obtained in different materials containing different levels of impurities. Also the temperature-independence is inconsistent with chemical reaction, for even at the diffusion-controlled limit $E_a \sim 17$ kJ mol^{-1} for Mu reactions in water (Ng *et al.*, 1981). (ii) For the same reasons λ_0 is unlikely to represent a chemical reaction with the solvent. In the case of water, the H-analogues of the reactions Mu + $H_2O \rightarrow$ MuH + OH or Mu + $H_2O \rightarrow$ H + MuOH are smaller than λ_0 by many orders of magnitude. Furthermore, reaction of Mu with water, neopentane, and N_2 is unlikely to be the same. (iii) Depolarization having some physical origin, such as dipolar or collisional broadening, again seems unlikely because of the observed independence on the nature of the medium. (iv) The possibility that magnetic field inhomogeneities, or drifts, cause some dephasing of the muon spins at long times cannot be ruled out. However, it seems improbable in view of the range of different magnets, power supplies and the magnetic fields applied. (v) λ_0 does not represent the reaction of Mu with radiation chemical products. There are two aspects here. First, the accumulated radiation dose produced in the medium by the high-energy muons at the end of a μSR experiment is only ~30 rad (0.3 Gy), creating $<2 \times 10^{-7}$ M in products (see chapter 5). This concentration is too low to account for $\lambda_0 = 2.5 \times 10^5 s^{-1}$ because diffusion-limited rate constants are $\sim 10^{10} M^{-1} s^{-1}$. Second, Mu cannot be observed to be disappearing by reaction with species created in its spur (or track) over 10^{-6}–10^{-5} s; because such non-homogeneous reactions are not kinetically first-order (exponential), and the local concentration falls to less than 10^{-6} M within a few microseconds due to diffusion (Walker, 1981).

All of the above applies to μSR studies in d.c. mode. However, exploiting the pulsed method and its very low background out to long times, Nagamine *et al.* (1982) found λ_0 in water in 2-3-G homogeneous fields to be $(0.5 \pm 0.2) \times 10^5$ s, five times smaller. The authors suggest that the difference arises from the fitting procedure: specifically that beating between the two muonium frequencies (ν_{12} and ν_{23} in Figure 3.5) can appear as a damping, just like a relaxation, and this depends on the fourth power of the magnetic field. Their much lower (and more homogeneous) field may then account for their lower – and therefore better – background decay. Indeed, this suggestion seems to be correct, because λ_0 can be reduced when a two-frequency fitting equation is used (Ng *et al.*, unpublished data).

As a final note it should be added that Mu reacts very efficiently with O_2, $k_M = 2.4 \times 10^{10} M^{-1} s^{-1}$ in water (Jean, Fleming, Ng & Walker, 1979), so that all media used for Mu studies need to have the oxygen concentration reduced to less than $\sim 2 \times 10^{-6}$ M. Since air-saturated water contains 2.6×10^{-4} M it is necessary to reduce this concentration by a factor of more than 120. This can readily be accomplished either by bubbling with an oxygen-free gas or by freeze-pump-thaw evacuation cycles; but trace amounts of O_2 are almost certainly not the origin of λ_0.

References for chapter 4

Brewer, J. H., Crowe, K. M., Johnson, R. F., Schenck, A. & Williams, R. W. (1971). Fast depolarization of positive muons in solution – the chemistry of atomic muonium. *Physical Review Letters*, **27**, 297–300.

Brewer, J. H. (1972). Muon depolarization and the chemistry of muonium in liquids. Ph.D. Thesis, University of California, Berkeley. *Dissertation Abstracts*, pp. 1–103.

Brewer, J. H., Gygax, F. N. & Fleming, D. G. (1973). Mechanism for μ^+ depolarization in liquids – muonium chemistry and radical formation. *Physical Review*, **A8**, 77–86.

Brewer, J. H., Crowe, K. M., Gygax, F. N., Johnson, R. F., Fleming, D. G. & Schenck, A. (1974). Muonium chemistry in liquids: evidence for transient radicals. *Physical Review*, **A9**, 495–507.

Brewer, J. H., Crowe, K. M., Gygax, F. N. & Schenck, A. (1975). Positive muons and muonium in matter. In *Muon Physics*, vol. 3, ed. V. W. Hughes & C. S. Wu, pp. 3–139. New York: Academic Press.

Fessenden, R. W. (1964). ESR studies of internal rotation in radicals. *Journal de Chimie physique*, **61**, 1570–5.

Fleming, D. G., Garner, D. M., Vaz, L. C., Walker, D. C., Brewer, J. H. & Crowe, K. M. (1979). Muonium chemistry – a review. *Advances in Chemistry Series*, **175**, 279–334.

Firsov, V. G. & Byakov, V. M. (1965). Chemical reactions involving muonium. A method for determining the absolute rate constants and other reaction parameters. *Soviet Physics JETP*, **20**, 719.

Garner, D. M. (1979). Application of the muonium spin rotation technique to a study of the gas phase chemical kinetics of muonium reactions with the halogens and hydrogen halides. Ph.D. Thesis, University of British Columbia, *Dissertation Abstracts*, 1–307.

Gurevich, I. I., Ivanter, I. G., Meleshko, E. A., Nikolskii, B. A., Roganov, V. S., Selivanov, V. I., Smilga, V. P., Sokolov, B. V. & Shestakov, V. D. (1971). Two-frequency precession of muonium in a magnetic field. *Soviet Physics JETP*, **33**, 253–9.

Ivanter, I. G. & Smilga, V. P. (1968). Theory of the muonium mechanism of depolarization of μ^+ mesons in media. *Soviet Physics JETP*, **29**, 301–6.

Jean, Y. C., Brewer, J. H., Fleming, D. G., Garner, D. M., Mikula, R. J., Vaz, L. C. & Walker, D. C. (1978). Reactivity of muonium atoms in aqueous solution. *Chemical Physics Letters*, **57**, 293–7.

Jean, Y. C., Fleming, D. G., Ng, B. W. & Walker, D. C. (1979). Reaction of muonium with O_2 in aqueous solution. *Chemical Physics Letters*, **66**, 187–90.

Miyake, Y., Tabata, Y., Ito, Y., Nishiyama, K. & Nagamine, K. (1982). To be published.

Nagamine, K., Nishiyama, K., Imazato, J., Nakayama, H., Yoshida, M., Sakai, Y., Sato, H. & Tominaga, T. (1982). Long-lived muonium in water revealed by pulsed muons. *Chemical Physics Letters*, **87**, 186–91.

Ng, B. W., Jean, Y. C., Ito, Y., Suzuki, T., Brewer, J. H., Fleming, D. G. & Walker, D. C. (1981). Diffusion and activation-controlled reactions of muonium in aqueous solutions. *Journal of Physical Chemistry*, **85**, 454–8.

Ng, B. W. (1980). *Studies of the Muonium Atom in Liquid Media*. M.Sc. Thesis, University of British Columbia.

Ng, B. W., Stadlbauer, J. M., Jean, Y. C. & Walker, D. C. (1982). Muonium atoms in liquid and solid neopentane. *Canadian Journal of Chemistry* (in press).

Percival, P. W. & Fischer, H. (1976). Theory and analysis of muon spin polarization in chemical systems. *Chemical Physics*, **16**, 89–99.

Percival, P. W., Roduner, E., Fischer, H., Camani, M., Gygax, F. N. & Schenck, A. (1977). Bimolecular rate constants for reactions of muonium in aqueous solutions. *Chemical Physics Letters*, **47**, 11–14.

Percival, P. W. (1981). (i) Use of spin polarized muons to probe hydrogen atom reactions. *15th International Free Radical Symposium, Nova Scotia.* (ii) *Annual Report of TRIUMF (1981)*, pp. 43–5. Vancouver: University of B.C.

Roduner, E., Percival, P. W., Fleming, D. G., Hochmann, J. & Fischer, H. (1978). Muonium-substituted transient radicals observed by muon spin rotation. *Chemical Physics Letters*, **57**, 37–40.

Roduner, E. (1979). *On the Liquid Phase Chemistry of the Light Hydrogen Isotope Muonium*. Ph.D. Thesis, University of Zurich, pp. 1–97.

Roduner, E. & Fischer, H. (1981). Muonium substituted organic free radicals in liquids. Theory and analysis of μSR spectra. *Chemical Physics*, **54**, 261–76.

Roduner, E. (1981). Muonium-substituted free radicals. *Hyperfine Interactions*, **8**, 561–70.
Schenck, A., Williams, R. W., Brewer, J. H., Crowe, K. M. & Johnson, R. F. (1972). Decay of the coherent spin precession of polarized positive muons in paramagnetic $MnCl_2$ solutions. *Chemical Physics Letters*, **12**, 544–8.
Schenck, A. (1976). On the application of polarized positive muons in solid state physics. In *Nuclear and Particle Physics at Intermediate Energies*, ed. J. B. Warren, pp. 159–297. New York: Plenum Press.
Stadlbauer, J. M., Ng, B. W., Walker, D. C., Jean, Y. C. & Ito, Y. (1981). Muonium addition to vinyl monomers. *Canadian Journal of Chemistry*, **59**, 3261–6.
Walker, D. C. (1981). Muonium: a light isotope of hydrogen. *Journal of Physical Chemistry*, **85**, 3960–71.

5

SOME COMPARISONS OF μSR WITH OTHER TECHNIQUES

μSR is an experimental method unique to the muon. It utilizes single-particle counting techniques in common with many nuclear physics studies; but its analogies with NMR and ESR are probably the first to come to mind, particularly to physicists, when comparing μSR with other techniques. The analogy of μSR with NMR arises, of course, from their common involvement with the interactions between nuclear moments, and with magnetic fields. Thus, μSR experiments in solid state physics use the muon's spin to probe local magnetic structures, dipolar interactions, Knight shifts, impurity centres, diffusion and trapping, and perhaps one day, chemical shifts. The kinship with ESR is even closer, because of the information to be gained from the muon's hyperfine interaction with unpaired electrons. This is particularly prominent at present because of the blooming of Mu-radical studies.

In general, however, chemical μSR is a technique for the study of short-lived species and their dynamics, whereas NMR and ESR are primarily used with relatively stable molecules to study their structure. So, the chemical kineticist finds muonium spin rotation studies to have more in common with pulse radiolysis studies of solvated electrons (e_s^-) and H-atoms, with positron annihilation studies of positronium (Ps), and with recoil atom studies of tritium (^{3}H).

Muonium is not 'only' an isotope of H, but has its own spot in the following series of unusual, highly reactive, light atoms or pseudo-atoms:

$$e_s^- - \text{Ps} - \text{Mu} - {}^1\text{H} - {}^3\text{H}$$

This series is of interest to kineticists and theorists concerned with the fundamental aspects of chemical reaction dynamics. It is also of interest to compare the techniques by which each of these species is studied.

Positronium is studied by three types of positron annihilation measurement: lifetime, angular correlation and Doppler broadening, with the lifetime method being the most commonly used for kinetic studies. Its rudiments are indicated in Figure 5.1. A histogram is accumulated from 10^6 to 10^8 individual positron lifetime measurements, and this is used to distinguish the chemical states and to calculate their reaction times. Clearly, the basic similarities with μSR are many. In fact, it is also only the triplet state of Ps whose decay can be observed. The singlet (*para*-Ps) is too short-lived (0.13 ns) for current technology, whereas

Figure 5.1(a). Schematic diagram of representative positron annihilation equipment required for lifetime measurements. (SCINT. is a fast plastic scintillator giving a light pulse from either the start (1.28 MeV) γ or from the stop (0.51 MeV annihilation) γ. The photomultipliers (PM) pick up the signal. A constant fraction discriminator (DISC) allows selection of the appropriate energy. These signals pass to a time-to-pulse-height-converter (TPHC) so that the observed lifetime is converted to an amplitude for binning in a multichannel analyser (MCA).)

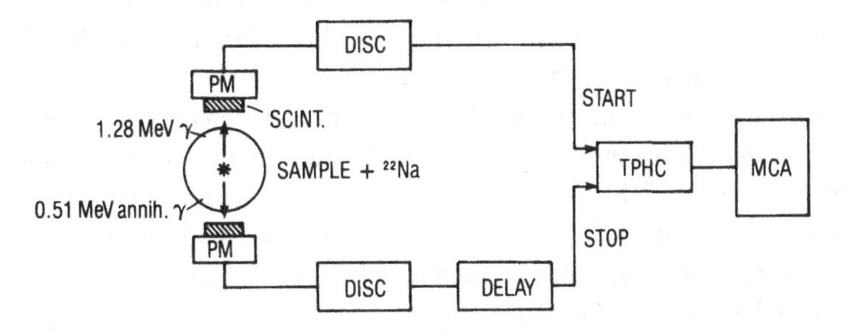

Figure 5.1(b). Typical lifetime histogram showing the computer fitting to a two-lifetime equation. (Figure following Green, 1972.)

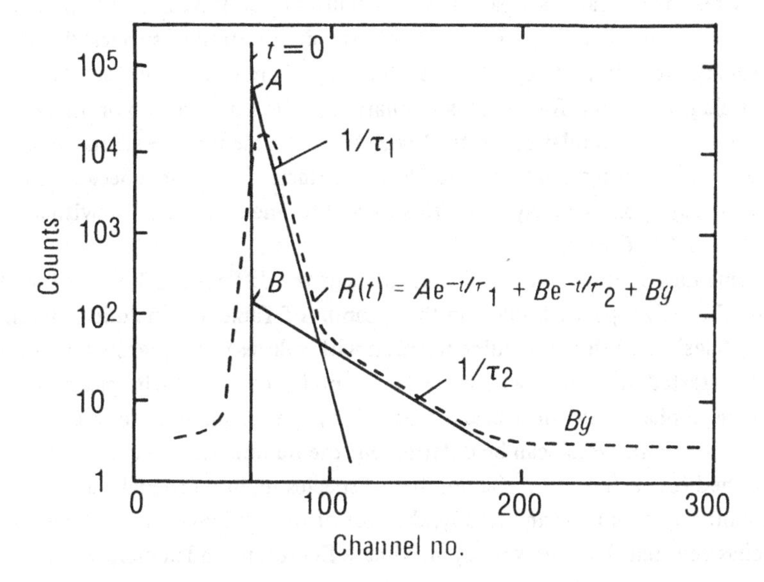

the triplet (*ortho*-Ps) has a lifetime of ~140 ns, *in vacuo*. In the presence of reactive compounds, Ps undergoes an interaction which changes the chemical state of the positron and thereby reduces its lifetime with respect to annihilation. In Mu, reaction also changes the chemical state - but this simply dephases the precession, without affecting the muon lifetime measurably. (The effect of chemical environment on the muon's natural decay rate is extremely small, as will be alluded to in chapter 11.)

Both e_s^- and H are most conveniently studied by pulse radiolysis: e_s^- studies are based on its optical absorption spectrum in the visible or near infrared, or its conductivity, and H studies are based on its characteristic ESR absorption. Four typical pulse radiolysis arrangements for optical studies are indicated by the sketch in Figure 5.2. For kinetic spectroscopy, an oscilloscope trace displays the absorption of light caused by the creation of a high concentration of e_{aq}^-, followed by their disappearance due to chemical reaction. Rate constants can be evaluated by analysis of the recovery rate of the transmitted light intensity.

Tritium is the only one of these five species not studied by a fast transient technique. Instead the reactivity of ^3H is deduced from the stable reaction products containing ^3H, and from the effect of temperature, pressure of moderator, and a range of scavengers, on their distribution. Such analyses are often accomplished using radiochromatography, so use is made of its radioactivity in determining where the tritium resides.

Each of these five species is chemically stable, over the normal observation period, in most common saturated hydrogenated solvents, such as water, alcohols and alkanes, or inert gases such as Ar, N_2 and H_2. But they are all highly reactive towards a great variety of solutes or additives. Anything reducible or unsaturated is readily attacked, so that the distribution of species' lifetimes reflects the reaction rate constant. Both e_s^- and H are intrinsically stable species; so would *o*-Ps be, but for its self-annihilation (140 ns in vacuum) or, more commonly, the annihilation of its e^+ with e^- of the medium (~1.3 ns in pure water). On the other hand, Mu and ^3H are intrinsically unstable because their nuclei decay spontaneously: μ^+ with a mean lifetime of 2.2 μs, $^3H^+$ with a half-life of 12.26 years.

Some idea about the relative versatilities of the different techniques used to study these five species is made in the columns of Table 5.1. Included therein are the species' stabilities (excluding reaction with solutes); the year in which the studies started in earnest; whether each method can be used with gases as well as condensed phases; the timescale monitored by the technique; the resulting range of rate constants which can be determined; the number of rate constants so far published for each species; the number of species required to make a typical rate constant measurement; and finally, the cost of the equipment used. It should be emphasized that these are very approximate figures, intended merely as a guideline. It is interesting to note the great flexibility in timescale of pulse radiolysis set-ups and the rather limited ranges of μSR and positron annihilation. Nevertheless, the 10^{-7} to 10^{-5} s scale of μSR is appropriate for studying reactions at the diffusion-controlled limit in the ideal concentration range $\sim 10^{-4}$ M. Only

Figure 5.2. Schematic diagram of four types of pulse-radiolysis arrangements for optical studies of transient species. (*a*) Fluorescence emission; (*b*) absorption spectrophotometry; (*c*) the stroboscopic interegative method of Hunt (1976); (*d*) Cerenkov reabsorption spectroscopy (Wallace & Walker, 1972). (S is the irradiated sample.)

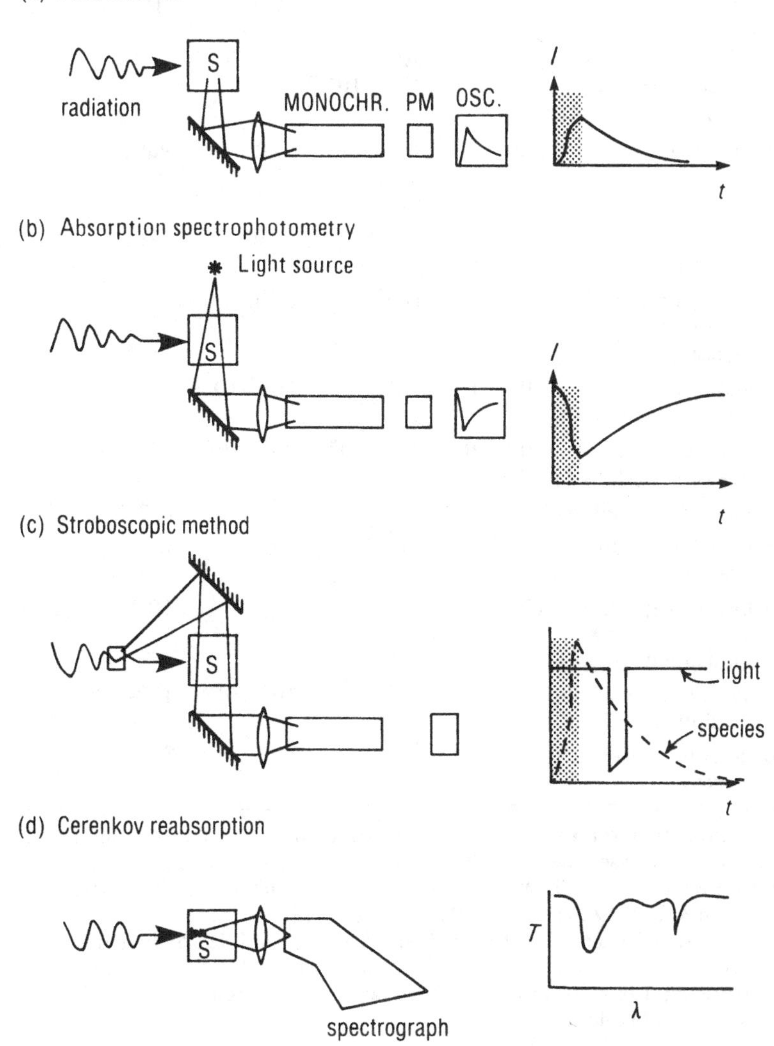

solute concentrations between 10^{-5} M and 1 M are considered in the Table, because below 10^{-5} M adventitious impurities are too important and above 1 M there are invariably ambiguities (see chapter 6).

It should be noted that these techniques generally show the actual rate of disappearance of the reactive species - except for the case of ^3H. Such measurements usually refer to the presence of a solute at relatively high concentration.

Table 5.1. *Comparison of some features of the techniques used to study the kinetics of e_s^-, Ps, Mu, ^1H and ^3H.*

Species	e_s^-	Ps	Mu	^1H	^3H
Stability of species (mean lifetime in inert medium (s))	∞	Annihilation (10^{-9})	2.2×10^{-6}	∞	5.6×10^{8}
Technique	Pulse radiolysis	Positron annihilation	μSR	Pulse radiolysis	Product analysis
Year measurements started in earnest	1962	1952	1960	1960	1950
Phases conveniently studied (c = condensed, g = gas)	c	c(+g)	c+g	c	c+g
Timescale of technique (s)	10^{-11}–1	10^{-10}–10^{-8}	10^{-7}–10^{-5}	10^{-7}–1	10^{2}–∞
Range of k values measured ($M^{-1}s^{-1}$)	10^{14}–10^{3}	10^{10}–10^{8}	10^{10}–10^{5}	10^{10}–10^{3}	N.A.
(solute conc. used/M)	(10^{-5}–1)	(10^{-2}–1)	(10^{-5}–1)	(10^{-5}–1)	–
Number of ks published	10^{3}	10^{2}	70	$10^{2.5}$	N.A.
Number of species required for a k measurement	10^{15}	10^{8}	10^{8}	10^{15}	10^{17}
Typical cost of whole equipment (£)	10^{5}–10^{6}	10^{4}	10^{7}–10^{8}	10^{5}–10^{6}	(10^{4}–10^{7})
Background references	[a]	[b]	[c]	[d]	[e]

The numbers given in this table are approximate and intended only as a guideline. In some cases they may be in error by an order of magnitude or so.
[a] Hart & Anbar, 1970; Walker, 1967; Anbar, Bambenek & Ross, 1973.
[b] Green & Lee, 1964; Green, 1972; Ache, 1979.
[c] Hughes, 1966; Percival, 1979; Firsov & Goldanskii, 1975; Fisher, 1981.
[d] Fessenden & Schuler, 1970; Anbar, Farhataziz & Ross, 1975.
[e] Wolfgang, 1965; Tominaga & Tachikawa, 1981; Urch, 1981.
N.A., not available.

The reactions are therefore bimolecular reactions studied under pseudo-first-order rate conditions. With regard to the Ps and Mu measurements, the products of their reactions are never identified. (The exception here is for Mu-radicals which are assumed to be already products of Mu addition reactions.) When studying e_s^- or H reactions, the immediate adducts or even final products can occasionally be observed and identified by optical absorption spectroscopy, conductivity, or ESR. But in many studies of all these species, one has to resort to chemical intuition, or guesswork, in order to write down the chemical reaction involved. With ^3H it is necessary to reverse the reasoning, and to infer the primary and intermediate reaction steps from the distribution of final stable molecules.

Mu, ^3H and Ps studies have in common the fact that an extremely rare charged particle is injected into the target material at very high energy; then, despite all the radiation chemical changes produced in the medium, the techniques merely focus on the original particle and observe its ultimate fate. This is in complete contrast with studies of e_s^- and H. These two are derived, or released, from the electrons and protons originally residing in the molecules of the medium, as a result of the application of a pulse of some form of high-energy radiation.

Also included in Table 5.1 is an indication of the absolute 'sensitivities' of these techniques, in terms of the number of species required in order to make a typical rate constant measurement. In no way do these numbers reflect the relative accuracy of the methods. Lifetime histograms in μSR and positron annihilation are usually composed of $\sim 10^7$ events. With a counting efficiency of ~10%, this means some 10^8 particles were injected into the medium, spread out over a period of some 10^2–10^5 s. In order to see the ESR absorption of H in micro-second pulse radiolysis it is necessary to have some 10^{15} H-atoms in the reaction cell. For e_s^- studies by optical absorption using radiation pulses as short as 10^{-11} s one needs to produce a comparable number of species. For instance, in order to get an optical absorbance of 10^{-2} in a 1-cm cube cell, even with the very strong absorptivity of e_s^- of $\sim 10^4$ close to the band maximum, one has to produce nearly 10^{15} e_s^- in the cell. Finally, for tritium studies, one needs enough ^3H to be able to measure several peaks corresponding to products containing ^3H for the quantitative analysis of a radiochromatogram. Suppose the limitation here is $\sim 10^{-6}$ g for each of 10 peaks of mean molecular weight 60, then some 10^{17} ^3H atoms had to be injected into a sample initially. If this was done *in situ* through either of the nuclear reactions $^3He(n,p)^3H$ or $^6Li(n,\alpha)^3H$, then the number of neutrons required would be some orders of magnitude larger still.

Finally, for this chapter, let us consider whether the radiation chemical changes induced by the energetic muon have a significant impact on μSR studies. The intention here is not to consider whether μ^+ and/or Mu actually emerge with excess kinetic energy from the last spur of the muon track and thermalize beyond it, because that is dealt with in the next chapter. The question posed here is the much simpler one, of whether the overall build-up of radiation products during the injection of 10^8 muons into a medium causes a progressive decrease in the muon's polarization or contributes to the background decay of Mu (λ_0). The mean molar concentration (c) of a stable product formed with a radiation yield G (mol/100 eV of energy absorbed) is given by Eq. [5.1],

$$c = GnE\rho/ar(6 \times 10^{22})\,\text{M} \qquad [5.1]$$

where n is the total number of particles of energy E (eV) deposited over an area a (cm^2) when the particles' mean range is r (cm) in a medium of density ρ (g cm^{-3}). Using values for a typical 'surface' μSR experiment in water: $n=10^8, E=4\times10^6$, $\rho=1, a=2$ cm^2 and $r=0.13$ cm, one obtains $c=G\times3\times10^{-8}$ M. G values for non-chain processes range from <0.1 to 5. Even with $G=5$, the concentration built up by the end of a μSR experiment having the parameters listed above is only $\sim1.5\times10^{-7}$ M. This mean bulk concentration is too small by at least two orders of magnitude to contribute to λ_0, because even with a rate constant at the diffusion-controlled limit ($k_M \sim 10^{10}$ M^{-1} s^{-1}) the ^{T}Mu chemical lifetime would be $\sim10^{-3}$ s - far beyond its spontaneous decay. It is also several orders of magnitude too small to cause significant primary depolarization of the muon.

The total dose deposited by the muons in a typical μSR experiment, using the above conditions, is only ~30 rad (0.3 J kg^{-1}). The point is, that although the muons form very densely ionizing tracks (Linear Energy Transfer ~ 30 MeV cm^{-1}) there are relatively few of them. Higher energy muon beams ('backward' or 'forward') give lower doses because their range-increase is greater than their energy-increase (percentagewise). The above estimate has not included the contaminants in the beam, however. With 'surface' muons these tend to be mainly decay-positrons with large ranges, therefore depositing small doses. Also, protons in the beam having the same momentum as the μ^+ (the magnets in the channel beamlines merely select particles by virtue of momentum and charge) have insufficient energy to penetrate the front wall of the reaction cell, so they too should not contribute appreciably to the total radiation dose.

It seems likely, therefore, that even by the end of a typical μSR experiment, the accumulated radiation damage to the sample will have an immeasurably small effect on either the yields or decay rates of the muon species being studied.

References for chapter 5

Ache, H. J. (1979). Positronium chemistry: present and future directions. *Advances in Chemistry Series* 175, 1–49. Washington: American Chemical Society.

Anbar, M., Bambenek, M. & Ross, A. B. (1973). Selected specific rates of reactions of transients from water in aqueous solution. 1. Hydrated electrons. *National Standard Reference Data System, N.B.S., No. 43.*

Anbar, M., Farhataziz & Ross, A. B. (1975). Selected specific rates of reactions of transients from water in aqueous solution. 2. Hydrogen atoms. *National Standard Reference Data Systems, N.B.S., No. 51.*

Fessenden, R. W. & Schuler, R. H. (1970). Electron spin resonance spectra of radiation-produced radicals. In *Advances in Radiation Chemistry*, vol. 2, ed. M. Burton & J. L. Magee, pp. 1-176. New York: Wiley-Interscience.

Firsov, V. G. & Goldanskii, V. I. (1975). Muonium. In *Radiochemistry*, vol. 8, ed. A. G. Maddock, MTP International Reviews of Science, Inorganic Chemistry, Series Two, pp. 1-47. London: Butterworth.

Fischer, H. (1981). Chemie mit myonen. *Chimia*, **35**, 85-94.

Green, J. & Lee, J. (1964). *Positronium Chemistry*. New York: Academic Press.

Green, J. H. (1972). Positronium and mesonic atoms. In *Radiochemistry*, ed. A. G. Maddock, MTP International Review of Science, Inorganic Chemistry, vol. 8, pp. 251-90. London: Butterworth.

Hart, E. J. & Anbar, M. (1970). *The Hydrated Electron*. New York: Wiley-Interscience.

Hughes, V. W. (1966). Muonium. *Annual Review of Nuclear Science*, **16**, 445-70.

Hunt, J. W. (1976). Early events in radiation chemistry. In *Advances in Radiation Chemistry*, vol. 5, ed. M. Button & J. L. Magee, pp. 185-315. New York: Wiley-Interscience.

Percival, P. W. (1979). Muonium chemistry. *Radiochimica Acta*, **26**, 1-14.

Tominaga, T. & Tachikawa, E. (1981). *Modern Hot-Atom Chemistry and Its Applications.* Berlin: Springer-Verlag.

Urch, D. S. (1981). Recent developments in gas-phase hot atom chemistry. *Radiochimica Acta*, **28**, 182-9.

Walker, D. C. (1967). The hydrated electron. *Quarterly Reviews*, **21**, 79-108.

Wallace, S. C. & Walker, D. C. (1972). Cerenkov reabsorption spectroscopy for subnano-second pulse radiolysis studies. *Journal of Physical Chemistry*, **76**, 3780-93.

Wolfgang, R. (1965). The hot atom chemistry of gas-phase systems. In *Progress in Reaction Kinetics*, vol. **3**, ed. G. Porter, pp. 97-169. London: Pergamon Press.

6

MUON REACTIVITY AND MUONIUM FORMATION

6.1 Preamble

From Swanson's (1958) initial measurements of the magnitude of the muon asymmetry (A_D, here) it was clear that the chemical fate of the muon varied dramatically with the composition of the medium. For instance, he found all the incident muons to be observable in diamagnetic environments in CCl_4, 62% of them in water and only 15% of them in benzene. Why this diversity, and what is the nature of the remaining muons? These questions occupy a good fraction of this chapter. When Hughes, McColm, Ziock & Prepost (1960) first observed free muonium atoms by direct μSR methods in argon gas, it became apparent that Mu atoms form and, at least in some systems, survive into the observation time-period of $>10^{-7}$s. Then when Mu-radicals were observed (Roduner *et al.*, 1978) the third type of identifiable muon magnetic state was finally established to exist with the necessary lifetime. But the sum of these three yields still does not account for the total number of incident muons in most systems, so there is also a missing fraction. Conceivably, this could be due to their existence in another type of magnetic state precessing in transverse fields at a frequency not yet seen; but, much more likely, the missing yield represents those muons which were spin-depolarized, or dephased, by fast interactions prior to the observation time of 10^{-7}s.

The intention here is to provide the basic information on these muon yields, and to report on the rationalizations that have been offered. The need to understand how, or why, various yields arise, stems not merely from insatiable curiosity, but the necessity to predict the effects in fresh systems and to plan studies into the unknown. Although many materials are mentioned in this discussion almost all yields published at present are collected alphabetically in Table A of the Appendix.

Muons are injected into a medium during a μSR experiment at very high kinetic energy (4.1 MeV for 'surface' muons and higher for a 'backward' beam).

They transfer this energy to the medium primarily by causing ionizations and molecular excitations of the molecules of the medium. During the last 10^4 eV or so of their tracks the muons are thought to undergo a sequence of charge-exchange cycles, interconverting between μ^+ and Mu as in: $\mu^+ \xrightarrow{e^-} \text{Mu} \xrightarrow{-e^-} \mu^+ \xrightarrow{e^-} \text{Mu} \xrightarrow{-e^-} \mu^+$ etc., in a series of electron pick-up and loss reactions represented by Eqs. [6.1] and [6.2]

$$\mu^+_{(\text{hot})} + \text{X} \xrightarrow{\sigma_{01}} \text{Mu}_{(\text{hot})} + \text{X}^+ \qquad [6.1]$$

$$\text{Mu}_{(\text{hot})} + \text{X} \xrightarrow{\sigma_{10}} \mu^+_{(\text{hot})} + e^-_{(\text{hot})} + \text{X} \qquad [6.2]$$

where X represents a molecule of the medium, σ_{01} is the pick-up cross-section of μ^+ and σ_{10} the cross-section for an ionizing collision of Mu. Each cycle consumes at least the ionization energy of X. The cross-sections are expected to change with kinetic energy and with the ionization energy of the medium – in the manner depicted in Figures 6.1 and 6.2 for the noble gases, by extrapolation from proton data assuming the Born approximation to hold. Except for He and Ne, whose ionization energies greatly exceed that of Mu (13.5 eV), the muons are expected to spend a high fraction of their time in the neutral (Mu) state over their last 10^3 eV of energy (see Figure 6.2). Whether they actually emerge at thermal energy in the μ^+ or Mu state depends on the collisional excitation, elastic and inelastic scattering of the final 20 eV or so.

But there are additional processes possible in molecular systems and particularly in the condensed phases. These are 'hot atom' reactions occurring by collision of $\text{Mu}^*_{(\text{hot})}$, and possibly by $\mu^+_{(\text{hot})}$, giving stable chemical products before these species reach thermal energy (~0.025 eV). By analogy with tritium recoil information, these hot-atom reactions of Mu* are likely to be abstractions, as in Eq. [6.3],

$$\text{Mu}^* + \text{RH (or RCl)} \rightarrow \text{MuH (or MuCl)} + \text{R} \qquad [6.3]$$

or substitution reactions, as in Eq. [6.4].

$$\text{Mu}^* + \text{RH (or RCl)} \rightarrow \text{MuR} + \text{H (or Cl)} \qquad [6.4]$$

For the most part they are likely to lead to the incorporation of the muon in a stable diamagnetic molecule. If no such epithermal reaction occurs, then the muon will eventually reach thermal energy either as a free μ^+ or as Mu. Its chemistry from that point will depend directly on the detailed chemical composition of the medium and will be governed by selective, competitive, thermal reactions. If the muon is still in touch with the trail of electrons, ions, free radicals and excited states of its own radiation track, then it may interact with one of them. If it is thermalized a considerable distance beyond this mélange of reactive species, then it can only react with the components of the bulk (homogeneous) medium, the solvent and its solutes. These alternative mechanisms are portrayed loosely by the scheme of Figure 6.3 and will be discussed in section 4 later. The period of time elapsing between the moment of thermalization in a liquid ($\sim 10^{-13}$ s) and that of observation by μSR ($>10^{-7}$ s) is eons on a chemical scale. These species can be solvated within 10^{-11} s, start to diffuse significantly within 10^{-9} s, and could have reacted with solutes present at 10^{-3} M by 10^{-7} s.

Figure 6.1. Charge-exchange total cross-sections for protons (top scale) and muons (lower scale) as calculated for He, Ne, Ar and Xe. Solid lines are the electron-capture cross-sections, dashed lines are the electron-loss cross-sections. (Figure taken from Brewer & Crowe, 1978.) The scale at the very bottom shows the equivalent positron energy (scaled according to mass) just to show the limited energy region over which positronium may be formed according to the ORE gap model (see section 5).

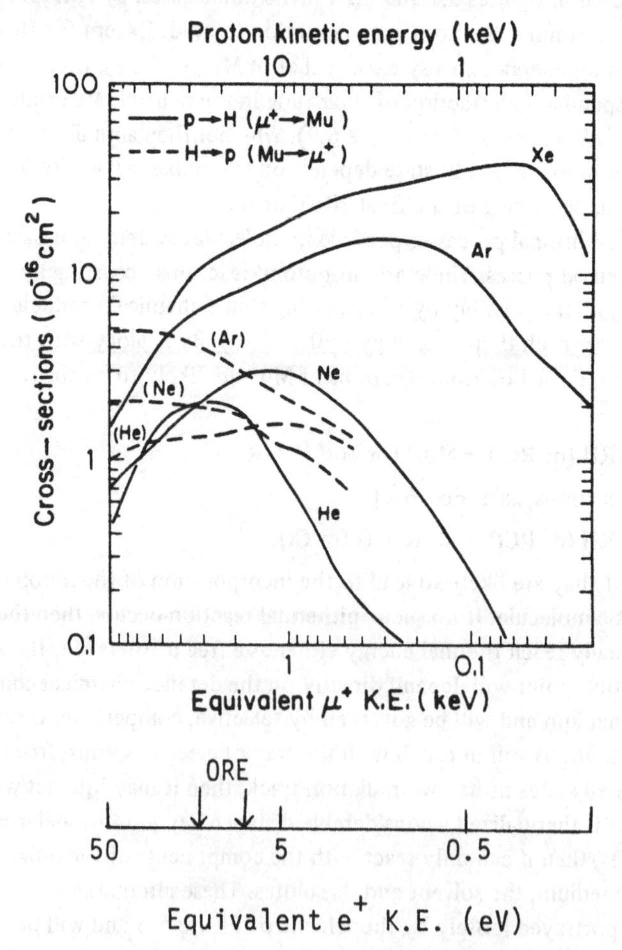

If hot (epithermal) reactions are largely responsible for the initial disbursement of muons into their observable states, then, it could be argued, one is not really studying muon chemistry in these systems. To do so properly, one would have to start with a thermalized μ^+; but this is currently beyond our reach, for while we need to use μSR we require polarized muons and therefore initially energetic ones. The chemistry of free μ^+ ions will have many features in common with the proton in chemistry (Bell, 1973). This may be less so in water where the proton (but not the muon) loses its identity; but even in rapid proton-exchange processes there could be a common thread. There may also be a strong isotope effect favouring the muon in many of its interactions.

6.2 Diamagnetic muon yields

In the absence of a method to distinguish different diamagnetic electron environments of the muon by μSR, it is necessary to lump them all together as

Figure 6.2. Plot showing the fraction of time spent in the neutral state (H or Mu) as a function of kinetic energy (K.E.) in various gases. The upper scale refers to H and the lower one to Mu. From the point of view of whether μ^+ or Mu finally emerges from the charge-exchange cycles one needs data further to the right than this plot goes; but it is clear that He is heading towards 100% μ^+, whereas Ar, N_2 and H_2 are heading towards 100% Mu. (Figure taken from Fleming *et al.*, 1979.)

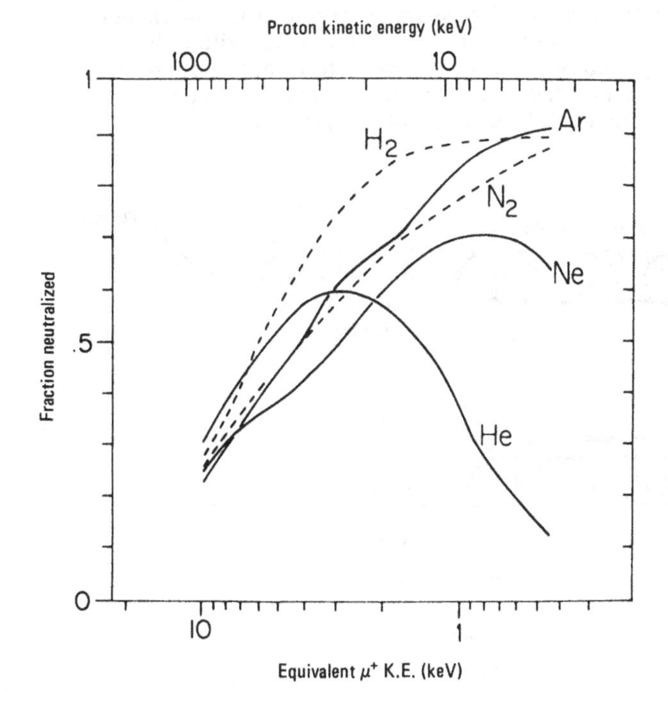

the fraction P_D. This usually prevents one deciding what the products are, let alone deducing the mechanism. For example, the diamagnetic fraction in water could arise from the formation of MuH through abstraction by hot Mu, or it could come from the presence of MuOH following thermalization of μ^+ and proton exchange. Not to be able to distinguish such fundamentally different processes at this stage is obviously quite limiting. Nevertheless, for many purposes it is not crucial to know the exact chemical nature of the diamagnetic states since they are all relatively inert once formed. It is the paramagnetic states that are most active and interesting, and μSR provides information on their electronic structure.

(a) *Pure materials*

The P_D values of some eighty pure chemicals have been determined. With the exception of CCl_4 and the noble gases, they tend to have 'intermediate' values of 0.15–0.85, rather than the extremes of 0 or 1.0. Thus, despite a variety of possible processes available, one particular process almost never completely dominates in any given system.

Figure 6.3. Oversimplified depiction of the main alternative models of muonium formation and muon reactivity. In the hot model (1) Mu* emerges from the dense track of spurs to form diamagnetic states (D) or thermalize as Mu. In the spur model (2), μ^+ stops within the terminal spur to undergo intraspur reactions with e^- to form Mu, or escapes to remain diamagnetic. Muonium-containing free radicals can be formed in unsaturated materials by addition of Mu either before or after thermalization.

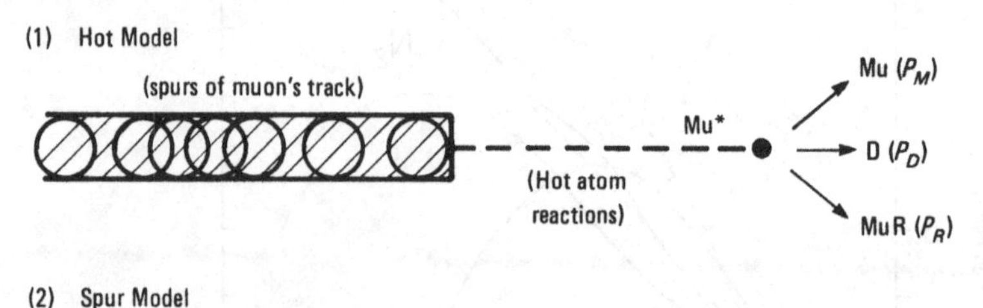

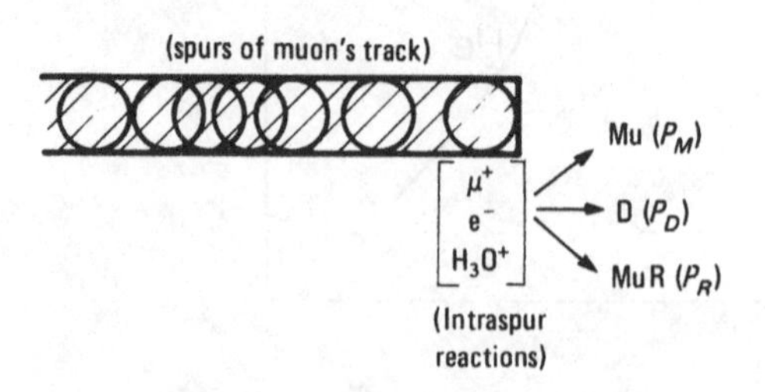

Noble gases. In the noble gases there is no possibility of hot-atom abstraction or substitution processes occurring and no secondary 'chemical' reactions are available. Therefore, they should be relatively simple, and that seems to be the case. Their P_D values change from 1.0 in He and Ne to 0 in Kr and Xe, commensurate with the change in ionization energy (see section 3 below). Neither are there any startling phase effects, as evident from Table 6.1. The diamagnetic component in these systems should be bare μ^+ (or weak molecular ions formed by association, such as $NeMu^+$ (Mikula, 1981)).

Table 6.1. *Selection of materials showing a full range of P_D values (see Appendix for a more complete list)*

Material	Phase	P_D	References
He	gas or liquid	1.0	(*a*), (*b*)
CCl_4	liquid	1.0	(*c*)
Al	metal	1.0	(*d*)
$CHCl_3$	liquid	0.85	(*c*)
glycerol	liquid	0.75	(*c*)
$(CHOH)_6$	solid	0.69	(*e*)
$c\text{-}C_6H_{12}$	liquid	0.68	(*c*)
$n\text{-}C_6H_{14}$	liquid	0.65	(*c*)
polyethylene	solid	0.64	(*f*)
H_2O	liquid	0.62	(*g*)
CH_3OH	liquid	0.62	(*g*)
$(CH_3)_2CHOH$	liquid	0.62	(*c*)
H_2O ice	solid	0.48	(*h*)
CH_3CN	liquid	0.43	(*e*)
C_6H_5OH	liquid	0.38	(*c*)
polystyrene	solid	0.31	(*f*)
$CH_2{=}CHCN$	liquid	0.28	(*e*)
fused quartz	solid	0.21	(*i*)
$n\text{-}C_6H_{14}$	gas	0.19	(*k*)
NaCl	solid	0.18	(*f*)
styrene	liquid	0.17	(*e*)
CS_2	liquid	0.16	(*c*)
C_6H_6	liquid	0.15	(*c*)
CsI	solid	0.14	(*f*)
Kr	liquid	0.07	(*j*)
Kr	gas	0	(*a*)

(*a*) Stambaugh *et al.*, 1974
(*b*) Crane *et al.*, 1974
(*c*) Fleming *et al.*, 1979
(*d*) Brewer *et al.*, 1975
(*e*) Stadlbauer *et al.*, 1982
(*f*) Swanson, 1958
(*g*) Percival *et al.*, 1978
(*h*) Percival *et al.*, 1979
(*i*) Brewer, 1981
(*j*) Keifl *et al.*, 1981
(*k*) Arseneau *et al.*, 1982

Saturated hydrogenated compounds. This group of materials, which includes water, alcohols, ethers and alkanes, all have the same diamagnetic yield in the liquid phase, $P_D = 0.64 \pm 0.05$ (see Table 6.1). Yet these compounds have totally different physical properties - particularly with regard to static dielectric constant and solvating power for ions in solution. This strongly implies that P_D does not arise from ionic processes such as the emergence of thermalized μ^+ from the thermalization track. Instead, it suggests that abstraction, or substitution, of H by hot Mu* proceeds with comparable efficiency relative to thermalization in all these saturated compounds.

Effect of phase. In sharp contrast to the liquid phase, all these saturated compounds seem to have P_D values very close to zero in the gas phase (Arseneau *et al.*, 1982). As in the heavy noble gases the absence of a diamagnetic component is associated with a 100% yield of muonium (see section 3 below). Perhaps it is not surprising that Mu is favoured over μ^+ in the charge-exchange cycles, since the ionization energy of these compounds is less than that of Mu. But why are there no hot-atom reactions of Mu* in these gases? Could it be that abstraction and substitution by such a light energetic particle requires the presence of a third body to stabilize the product and carry-off the exothermicity? Perhaps thermalization of Mu* wins under the rarefied conditions of gases at atmospheric pressure.

Other phase effects of note in Table 6.1 include the decrease in P_D upon freezing water. But this is the only material showing a significant change in P_D at the liquid-solid transition, and it may result from the structural alteration and enhanced hydrogen-bonding. Neopentane shows no phase effect, and liquid hexane corresponds closely with that of solid polyethylene.

Bond energy. Consideration has to be given to the possibility that P_D increases as the strength of a bond broken decreases in the compound studied. This would be expected, generally, if P_D arose from selective thermal reactions of Mu which were activation-controlled. On the other hand, there should be no strong influence if hot Mu* reactions are responsible. Care has to be exercised in selecting compounds to examine in this regard. First, most molecules contain more than one type of chemical bond, and it is not known which bond is broken. Second, the thermalization of Mu* occurs more efficiently during collisions with lighter atoms, so that the total number of collisions made during thermalization changes with chemical composition. It would not be meaningful to attempt to compare CCl_4 with C_6H_6, for instance, on the basis of mean, or weakest, bond energies because both the above factors are involved. However, CCl_4 may usefully be compared with other tetrahedral tetrachlorides in which only the size and mass of the central atom changes.

Table 6.2 gives the P_D values for the Group IV series CCl_4, $SiCl_4$ and $SnCl_4$, from which P_D is seen to follow the inverse of the bond energies - with the strongly-bonded $SiCl_4$ having a small P_D. But any simple correlation here seems to be broken by the case of $TiCl_4$. Here the bond strength is even higher than in $SiCl_4$, yet P_D is 1.0. There may, of course, be other factors which can explain why $TiCl_4$ is high; but, equally, there may be other reasons why $SiCl_4$ is low. Another example with some merit is the comparison of 1-propanol with 1-propanol. They both have the same P_D yet 2-propanol contains a substantially weaker bond than 1-propanol.

Effect of π-bonding. A striking correlation exists between the value of P_D and the degree of π-bonding and its delocalization in a compound. First, P_D changes progressively with the degree of unsaturation. This can be seen in the data of Table 6.3 for three series: 2-propanol (0.62) and acetone (0.54); *c*-hexane (0.68), *c*-hexene (0.47), *c*-hexadiene (0.40) and benzene (0.15); and hexane (0.62), hexene (0.50) and hexyne (0.43). It appears that one triple bond has a

Table 6.2. *Effect of bond energy on P_D*

Compound	Bond energy (kJ mol^{-1})	P_D[a]
CCl_4	326	1.0
$SiCl_4$	380	0.48
$SnCl_4$	318	0.99
$TiCl_4$	426	1.0

[a] Data taken from Fleming *et al.*, 1979.

Table 6.3. *Effect of π-bonding on P_D in liquids*

Compound	P_D[a]
n-hexane	0.65
n-hexene	0.50
n-hexyne	0.43
c-hexane	0.68
c-hexene	0.47
1.4-*c*-hexadiene	0.40
1.3-*c*-hexadiene	0.32
benzene	0.15
2-propanol	0.62
acetone	0.54

[a] Data taken from Fleming *et al.*, 1979.

comparable effect to two non-conjugated double bonds. Second, P_D decreases more when there is conjugation of the double bonds, as in 1.3-*c*-hexadiene (0.32) compared to 1.4-*c*-hexadiene (0.40).

Perhaps the relatively low-lying electronic energy levels resulting from π-delocalization enhance the efficiency of thermalization of Mu*, thereby reducing the probability of a hot-atom abstraction or substitution reaction. Hot-atom addition reactions may be relatively unimportant due to the need to dispose of the initial kinetic energy; but thermalized Mu should readily add to π-bonds to form Mu-radicals.

A comparison with radiation-chemical yields is called for here. In the radiolysis of *c*-hexane, *c*-hexene and benzene, the *G*-values for H_2 production are 5.6, 1.3 and 0.039 mol/100 eV, respectively (Spinks & Woods, 1976). Other products such as total C_{12} compounds show a similar marked reduction with the extent of unsaturation in the irradiated materials. Not only is the overall decomposition reduced, but a product like H_2, which is believed to result partly from hot-atom reactions in hydrocarbons (Spinks & Woods, 1976), is dramatically curtailed by π-bonding and extensive delocalization. This analogy adds credence to the general notion that hot-atom processes, such as abstraction, are largely responsible for the diamagnetic products in μSR.

Effect of substitution by halogens or OH. Table 6.4 provides clear evidence that substitution of H by a halogen increases P_D. It increases with the extent of

Table 6.4. *Effect of halogenation on P_D in liquids*

Compound	P_D [a]
C_nH_{2n+2}	0.65 ± 0.3
CH_2Cl_2	0.70
$CHCl_3$	0.85
CCl_4	1.0
$CHBr_3$	0.88
C_6H_6	0.15
C_6H_5Cl	0.23
C_6H_5Br	0.38
C_6H_5I	0.49
$C_6H_5CH_2Cl$	0.35
$C_6H_5CHCl_2$	0.46
$C_6H_5CCl_3$	0.65 [b]
CCl_4 (vapour)	0.5 [c]

[a] Data taken from Fleming *et al.*, 1979.
[b] Stadlbauer *et al.*, 1982.
[c] At ~1 atmosphere (101 kN m^{-2}) pressure, Arseneau *et al.*, 1982.

substitution and with the mass of the halogen. It is evident in aliphatics, aromatics and in substituted toluenes. Data for the chlorinated methanes are presented in graphical form in Figure 6.4. This plot suggests that $P_D = 1.0$ is 'only just reached' by CCl_4. This inference is supported by the fact that when CCl_4 is 'diluted' by benzene (or cyclohexane) the P_D value falls nearly linearly with volume composition at the CCl_4 end – see Figure 6.5. It is also of interest to note that the P_D value of $C_6H_5CCl_3$ (0.65) is almost exactly the same as a 50:50 mixture of benzene and CCl_4 (see Figure 6.5).

These results imply that diamagnetic states do not result from formation mechanisms involving selective, competitive reactions of thermalized intermediates. They are consistent with non-competitive hot-atom abstraction or substitution reactions in which the efficiency towards Cl greatly exceeds that towards H. The increase in the series Cl → Br → I could arise from the higher electron density, the larger atomic volume, or the weaker bond energy.

Substitution of H by OH has very much less clear-cut effects on P_D than is the case with the halogens. There is a strong effect in the aromatics – toluene (0.37) compared to benzene (0.15) – but a rather uncertain effect in the aliphatics, with glycerol well up (0.75) while inositol, $(CHOH)_6$, is only marginally up (0.69).

Figure 6.4. Variation of P_D with extent of substitution of H by Cl. Circles represent CH_2Cl_2, $CHCl_3$ and CCl_4; triangle corresponds to saturated hydrocarbons, such as *n*-hexane. (Data from Fleming *et al.*, 1979.)

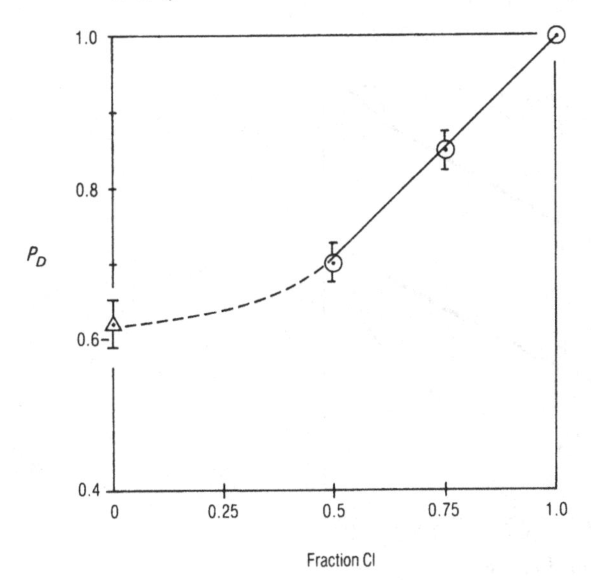

Figure 6.5. (a) Values of P_D obtained for mixtures of benzene with CCl_4 or cyclohexane plotted as a function of the volume-fraction of benzene. Circles refer to the left-hand side positron detector, triangles to the right-hand side detector. Statistical errors are comparable to the size of the data points, and random absolute uncertainties are <10%. (b) Curves **1** and **2** are the CCl_4 and cyclohexane plots of (a) above. Curve **3** is expected on the basis of competitive thermal reactions such as intraspur electron scavenging by CCl_4. Curve **4** is expected if benzene can 'protect' cyclohexane by intermolecular energy transfer or Mu scavenging to give radicals. (Figures taken from Jean *et al.*, 1981.)

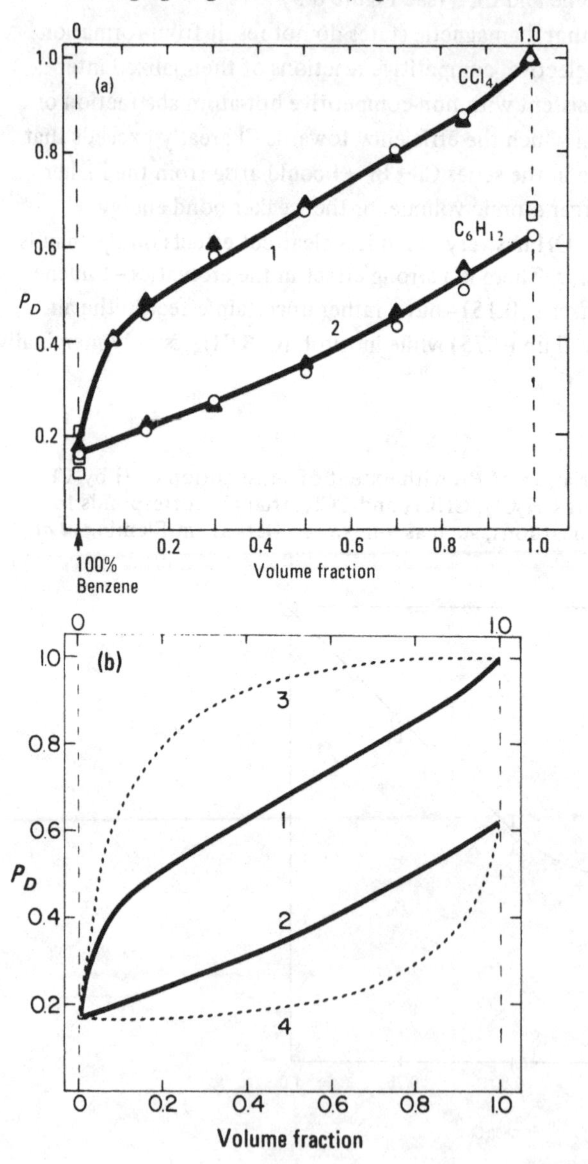

Decay of diamagnetic species. There is no observable decay of the diamagnetic signal ($\lambda_D \leqslant 10^3 s^{-1}$) over the μSR time-window, except in very rare cases, one of which is CS_2 (Ng, 1980). Perhaps this is not surprising. In general, diamagnetic species do not react under thermal conditions with diamagnetic molecules to give paramagnetic states: and conversions from one diamagnetic state to another will not be registered by μSR since they involve immeasurably small changes in precession frequencies. Consequently, it is only the *yields* of the diamagnetic states that are of interest. And as seen above, these cover the whole range 0 to 1.0, with the extremes as exceptions. They are strongly influenced by chemical composition, but relatively little by physical properties - being particularly insensitive to properties governing ionic processes.

(b) *Mixtures of solvents*

Having seen that pure CCl_4 and benzene provide extreme values of P_D for polyatomic liquids, it is natural to wonder what happens in their mixtures. The results are shown in Figure 6.5. There is a sharp increase in P_D caused by the presence of a small amount of CCl_4 in benzene (as there is in CCl_4-cyclohexane mixtures too (Jean *et al.*, 1982)); but over the rest of the composition range P_D changes almost linearly with volume fraction. The muon's fate seems to be determined solely by the fraction of time it spends in contact with the molecules of each kind. This was interpreted (Jean *et al.*, 1981) as an indication of intramolecular hot reactions, as if the cross-section for Mu* reactions peaked over an energy range comparable to the energy loss per encounter. The fate of the muon was sealed while in contact with only one molecule, without the chance to discriminate between the two components.

Other solvent mixtures have been studied over the years (Babaev *et al.*, 1966; Myasishcheva, Obukha, Roganov & Firsov, 1967; Brewer, Crowe, Gygax & Schenck, 1975) including methanol-chloroform and methanol-benzene as in Figure 6.6. In general they too show an approximately linear relation when P_D is plotted against volume fraction. There is the danger, of course, that some of these combinations form imperfect mixtures in which there are fairly large regions consisting of just one component. Linear plots would be expected under those conditions regardless of the mechanism. However, the benzene–cyclohexane mixture shown in Figure 6.5 is believed to be truly homogeneous and it too shows a linear change in P_D. This result implies that benzene affords no 'protection' of cyclohexane, in complete contrast to the H_2 yields in the radiolysis of these mixtures - where curve **4** of Figure 6.5 is obtained (Manion & Burton, 1952). This general absence of intermolecular protection, either as an energy sink or by sacrificial decomposition, as occur in radiolysis, again leads one to the conclusion that the diamagnetic muon states are formed by direct, one-step, non-competitive processes, with no intermolecular energy-transfer.

(c) *Solutions*

The only indication that thermalized muonium atoms may be converted to diamagnetic species in the mixtures of organic liquids discussed above, is in the presence of a trace of CCl_4 in benzene or cyclohexane. However,

Figure 6.6. P_D data (measured μ^+ asymmetry on left and residual polarization $\equiv P_D$ on right) plotted against volume-fraction for (a) mixtures of CH_3OH in C_6H_6, and (b) $CHCl_3$ in CH_3OH. The dotted curves indicate the expectations based on scavenging by $CHCl_3$ in (b) and protection by benzene in (a). (Figures taken from Fleming *et al.*, 1979.)

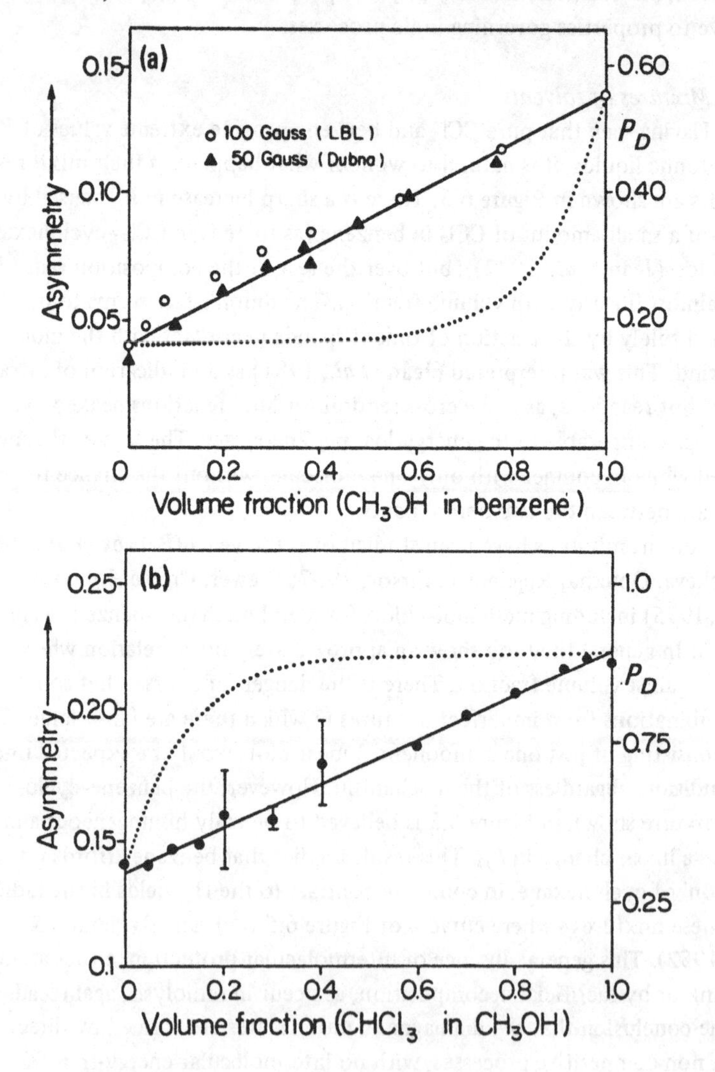

there are many solutes which convert Mu → D in water when present in high enough concentration to effect the conversion before significant dephasing of the muon in Mu has occurred. An example leading to nearly complete conservation of spin polarization at 1 M concentration is provided by the data for $S_2O_3^{2-}$ in Figure 4.4.

Whether or not a solute enhances P_D depends on the type of reaction and its rate. This is demonstrated by the range of solutes given in Table 6.5. Both oxidation and abstraction (or substitution) increase P_D by converting Mu → D. But spin-exchange does not, because it merely depolarizes the muon by ^{T}Mu → ^{S}Mu conversion, nor does addition to form a Mu-radical. In the presence of OH^- the equivalent of a proton transfer occurs, as well as a possible alteration of the muon's primary charge-exchange cycles due to $\mu^+ + OH^- \rightarrow$ MuOH. No effect was found with HCl up to 5 M. Two odd results are also reported in Table 6.5 to remind one of possible difficulties. With Ag^+, there was not the enhancement of P_D expected to arise from the conversion Mu → μ^+ (as if the intermediate, $AgMu^+$, was sufficiently long-lived to depolarize the muon). With ClO_4^-, which does not react significantly with Mu, there was an unexpected enhancement of P_D. Perhaps this latter is an example of a direct hot-atom interaction with a solute.

Table 6.5. *Effect of very high concentrations of solutes on P_D in water at 50–100 G*

Solute	Conc. (M)	P_D	Type of reaction	Conversion	Reference
0	0	0.62	–	–	(*a*)
CrO_4^{2-}	1	0.98	Oxidation	Mu → μ^+	(*b*)
H_2O_2	5	0.9	Abstraction (or substitution)	Mu → D	(*c*)
OH^-	10	0.75	Proton transfer	Mu → MuHO + e_s^-	(*d*)
Ni^{2+}	1	0.62	Spin-exchange	^{T}Mu → ^{S}Mu	(*d*)
$(CHCOOH)_2$	1	0.62	Addition	Mu → Mu-rad	(*d*)
Cd^{2+}	1	0.64	(Reacts with e^- only)	–	(*e*)
HCl	5	0.62	No reaction	–	(*a*)
Ag^+	1	0.62	(Oxidation?)	odd result	(*d*)
ClO_4^-	4	0.85	(No reaction)	odd result	(*d*)

(*a*) Percival *et al.*, 1976, 1978.
(*b*) Percival, 1981.
(*c*) Brewer *et al.*, 1975.
(*d*) Jean *et al.*, 1982.
(*e*) Walker *et al.*, 1979, and Percival *et al.*, 1978, 1979.

6.3 Muonium yields

Free muonium atoms have been observed by μSR only in a limited number of media: these include the heavier noble gases, most fully saturated hydrogenated compounds (H_2O, NH_3, CH_3OH, $(CH_3)_2O$ and alkanes), some inert gases such as N_2 and H_2, and a variety of non-metallic solids, particularly oxides. Mu has not been found in any compounds containing halogens or π-bonds, or in any system except when highly purified and deoxygenated. It was initially observed in gases (Hughes *et al.*, 1960), then in solids (Gurevich *et al.*, 1971), and finally in liquids (Percival *et al.*, 1976). Its yield varies considerably. It has been observed over a broad range of temperatures, and its yield sometimes changes abruptly at a phase transition.

Table 6.6. *P_D, P_M, P_L and P_R values in various materials and in different phases*

Group	Compound	Phase	T (K)	P_D	P_M	P_L	P_R	Reference
Noble gases	He	gas	295	1.0	0	0	0	(a)
		liquid	4	>0.9	0	0	0	(b)
	Ar	gas	295	0.25	0.75	0	0	(c)
		liquid	85	0.02	0.48	0.50	0	(d)
		solid	77	0.008	0.91	0.08	0	(d)
	Kr	gas	295	0	1.0	0	0	(a)
		liquid	120	0.07	0.57	0.36	0	(d)
		solid	90	0.01	0.71	0.28	0	(d)
Water	H_2O	gas	420	0.07	0.93	0	0	(e)
		liquid	295	0.62	0.20	0.18	0	(f)
		solid	270	0.48	0.52	0	0	(f)
	D_2O	liquid	295	0.57	0.18	0.25	0	(f)
		solid	270	0.39	0.63	0	0	(f)
Hydrocarbons	CH_4	gas	295	0.12	0.88	0	0	(e)
	$C(CH_3)_4$	liquid	295	0.59	0.19	0.22	0	(g)
		solid	245	0.63	0.18	0.19	0	(g)
	C_6H_6	liquid	295	0.15	0	0.20	0.65	(i)
	$C_6H_5CH{=}CH_2$	liquid	295	0.25	0	--0.75--		(i)
Miscellaneous	CCl_4	gas	350	0.5	0	0.5	0	(e)
		liquid	295	1.0	0	0	0	(h)
	N_2	gas	295	0.16	0.84	0	0	(h)
	H_2	gas	295	0.4	0.6	0	0	(e)
	Ne + 0.15%Xe	gas	295	0.19	0.81	0	0	(h)
	C_2H_5OH	liquid	295	0.59	0.20	0.21	0	(f)
	CS_2	liquid	295	0.16	0	0.84	0	(j)
	CH_3CN	liquid	295	0.43	0	0.57	0	(i)

(a) Stambaugh *et al.*, 1974
(b) Crane *et al.*, 1974
(c) Barnett *et al.*, 1975
(d) Keifl *et al.*, 1981
(e) Arseneau *et al.*, 1982
(f) Percival *et al.*, 1978, 1979
(g) Ng *et al.*, 1982
(h) Fleming *et al.*, 1979
(i) Stadlbauer *et al.*, 1982
(j) Ng, 1980

Noble gases. Table 6.6 provides some idea of the range of the fractional Mu yields, P_M, for the noble gases (the complete list appears in the Appendix). There is a clear transition from P_D to P_M as the ionization energy of the gas approaches that of Mu. The species which emerge from the track seem to be determined simply by the energetics of the charge-exchange sequences, as indicated by Figure 6.2. For He and Ne there is an energy below which reaction [6.1] becomes endoergic and Mu can no longer form significantly. For the heavier noble gases, reaction [6.1] wins at the end of the track, *where it counts*, with the result that $P_M = 1.0$. A similar pattern prevails in the condensed phases, again with the sharp transition evidently correlated to the difference in ionization energies.

No hot-atom reactions are possible in these noble monatomic gases, so the muons have only the options of thermalizing as Mu or μ^+, as determined by charge-exchange. It is significant that in Ar, where both Mu and μ^+ appear, there can be no interconversion, because the yields observed at $\sim 10^{-6}$ s represent the in-phase species formed at $\sim 10^{-9}$ s. Dilute mixtures of Xe in Ne, for instance, also show both Mu and μ^+ with $P_M + P_D \sim 1$ (Fleming *et al.*, 1979) so that interconversion by the thermal equivalent of reaction [6.1] evidently does not occur once the muon has been thermalized.

Water. Table 6.6 provides the P_M, P_L and P_D value for H_2O and D_2O in various phases. No Mu radicals have been detected in water. It is significant that P_M is much larger in the ices and larger in D_2O than in H_2O. P_L is also larger in D_2O than H_2O liquids, but is zero in both ices. There is a small temperature variation in addition to the abrupt change at the melting (and boiling) point in both P_M and P_D as shown in Figure 6.7. Water vapour at atmospheric pressure and $\sim$380 K gives $P_M \sim 1$, as do many inert gaseous materials whose ionization energy is less than the 13.5 eV of Mu. Many of these results will be discussed later in connection with the dichotomy of the hot and spur models (Figure 6.3).

Saturated hydrogenated compounds. One of the most prominent effects observed in μSR studies of liquids is that for all the saturated hydrogenated compounds studied - water, alcohols, and alkanes - the yields are virtually the same. Values of P_M, P_L and P_D are given in Table 6.6 or the Appendix. Since these liquids show an enormous variation in physical properties - particularly with respect to solvating power - their uniform yields must result from common chemical characteristics, probably one of which is having all valence fully saturated with hydrogens. This suggests that hot-atom abstraction or substitution of H by Mu* is responsible for the diamagnetic yield, and that similar thermalization efficiencies, dominated by collisions with C- or O-bonded

H-atoms, determine the fraction of Mu which reaches thermal energy uncombined.

The contrast with radiolysis is again of value. In the spurs and ionization tracks of high-energy particles, alkanes produce large quantities of H_2 ($G \sim 3$–5) but miniscule free ion yields ($G \sim 0.03$). These yields are attributed to the low dielectric constants of the media and the high mobility of free electrons, which almost complete geminate neutralization of the cations by the free electrons. In complete contrast, water has a smallish H_2 yield ($G \sim 0.45$) but large free-ion yield ($G \sim 2.9$). This is attributed to its high dielectric constant, strong solvating power, and relatively low-mobility electrons.

The similarity of their muonium formation efficiencies and the diversity of their radiolysis electron survival probabilities is displayed in Table 6.7 for water, methanol, *c*-hexane and neopentane. Evidently, there is no correlation between P_D and the radiolysis free-ion yield, nor between P_M and the geminate-ion yield. This marked contrast necessarily implies that muonium is not formed by intra-spur electron combination reactions.

Figure 6.7. Variation of P_D and P_M in H_2O with temperature and phase. Circles, P_D values from Percival *et al.* (1978); triangles, PD values from Ng, 1980; squares, P_M values from Percival *et al.* (1978); crosses, P_M values from Ng, 1980. In the gas phase $P_D \rightarrow 0$ and $P_M \rightarrow 1.0$ (Arseneau *et al.*, 1982). Solid lines connect P_M and P_D data.

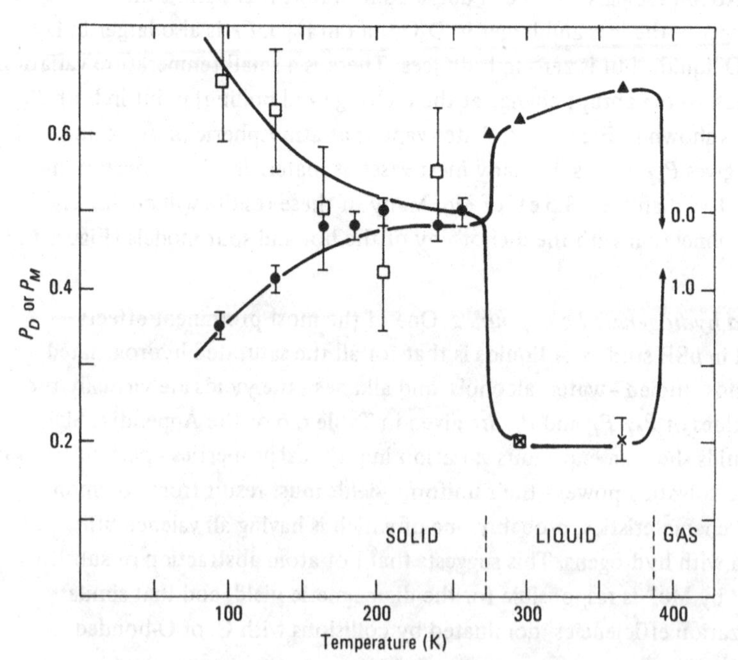

These conclusions are corroborated by the results of some μSR experiments that have been performed with a large electric field gradient applied to the liquid under study (Ito, Ng, Jean & Walker, 1981). An indication of the equipment used is sketched in Figure 6.8, showing the electric field applied in the direction of the incoming muon beam, perpendicular to the magnetic field, and with decay-positrons detected to right or left. Even at fields up to 20 kV cm^{-1} there was no observable change in P_D value in any of the liquids studied (n-hexane, c-hexane, benzene, CS_2 and CCl_4). The results were analysed in terms of Eq. [6.5],

$$p(e) = \exp(-r_c/r_0)[1 + r_c E/2\bar{k}T] \qquad [6.5]$$

where $p(e)$ is the escape probability of an ion-pair separated by distance r_0, in a medium whose Onsager critical escape distance is $r_c = e^2/\epsilon\bar{k}T$, and whose dielectric constant is ϵ at temperature T, with $\bar{k}$ Boltzmann's constant, and E the electric field gradient. The absence of an observable change in P_D (<3%) meant that if Mu arises from μ^+ combining with e$^-$ then their initial separation must have been less than 3.5 nm. This is much smaller than r_c in alkanes which is typically ~30 nm. In similar radiation chemical experiments, field gradients of ~10 kV cm^{-1} increased the intraspur electron escape probability considerably (Allen, 1976*a*), and in positron annihilation studies such fields reduced positronium formation (Mogensen, 1975). So muonium apparently is not formed in analogous intraspur processes - at least not with comparable initial charge-separation distances.

Table 6.7. *Comparison of P_D with the radiolysis free-ion yields in four types of pure liquids in which Mu is observed; the physical properties of static dielectric constant (D_s) and electron mobility (μ) are also given*

Compound	P_D	P_M[(a)]	$G(fi)$[(b)]	$P_D/(1-P_D)$[(c)]	$G(fi)/G(gi)$[(d)]	D_s	μ[(e)]
H_2O	0.62	0.2	2.8	1.6	1.6	78	0.002
CH_3OH	0.62	0.23	1.7	1.6	0.61	33	0.0006
c-C_6H_{12}	0.68	0.20	0.03	2.1	0.007	1.8	0.4
$C(CH_3)_4$	0.60	0.18	0.10	1.5	0.02	1.8	70

In these liquids the remaining muon yield, $1-P_D$, is probably equal to the initial Mu yield - see text. Therefore the ratio $P_D/(1-P_D)$ is used for comparison with the free-ion yield, $G(fi)$, divided by the geminate-ion yield, $G(gi)$, as determined by radiolysis for low LET radiations.

(a) P_M data are taken from Ito *et al.*, 1980, and Ng *et al.*, 1982.

(b) $G(fi)$ data from Allen, 1976*a*.

(c) Had P_D/P_M been used there would have been an equally small variation down the Table.

(d) $G(fi) + G(gi)$ taken to be equal to 4.5.

(e) Data from Allen, 1976*b*.

Unsaturated hydrocarbons. As seen in Table 6.3 and discussed in section 2 above, P_D is strongly reduced by the presence of π-bonding in the molecules of the medium. If this reduction results from more efficient thermalization of Mu*, rather than hot-atom addition to the π-bond, then the *initial* Mu yield must be higher in the unsaturated compounds than in saturated ones. However, no muonium is observed by μSR at $\sim 10^{-6}$ s in any unsaturated compound. So thermal Mu addition evidently occurs rapidly, and indeed, Mu-radicals are seen in most cases.

Miscellaneous compounds and effect of phase. No muonium is found in liquid CCl_4, in accord with $P_D = 1.0$; but in the vapour phase both P_M and P_D are about 0.5 (Arsenau *et al.*, 1982). For inert gases like N_2 and H_2 some 60-80% of

Figure 6.8. Schematic diagram of the orientation used for μSR experiments with an applied external electric field. The surface μ^+ beam is collimated by a Pb shield, triggers a muon 'start' counter C, enters and stops in the sample S placed in a transverse magnetic field created by the Helmholtz coils H. The muon lifetime is determined from the 'stop' signal received by the decay-positron telescope, L_1, L_2 and L_3. The thin cell has a pair of gauze electrodes E_1 and E_2 to which voltages up to 10 kV can be applied by power supply V. The direction of the applied electric field can be changed either with or opposed to the muon direction.

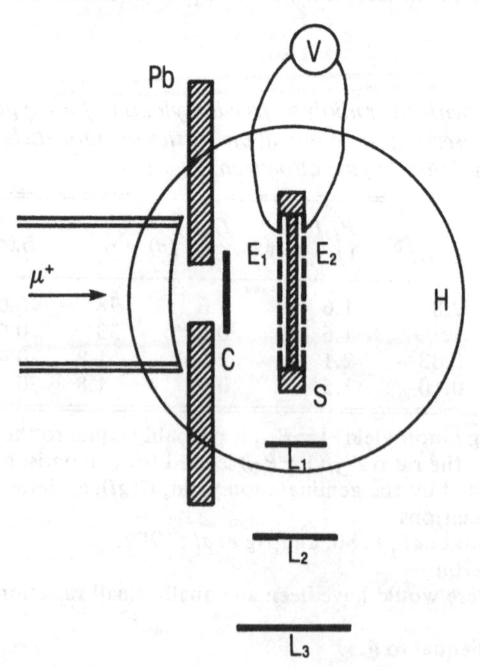

the muons are observed as Mu. In vapours of saturated alcohols and alkanes the muonium yield is close to 80%, suggesting that thermalization wins out over hot-atom abstraction in gases, in contrast to the condensed phases. Whether such yields are independent of pressure at very high pressures remains to be seen.

At the liquid–solid phase transition the changes in yields are less dramatic but quite interesting nevertheless. A selection of data is put together in Table 6.8. Everything changes in water, but not in ammonia or neopentane. But the freezing of water causes a considerable change in the hydrogen-bonded structure, whereas neopentane merely becomes glassy – probably with virtually the same structure and ordering as the liquid, only of very high viscosity. In xenon there seems to be a complete shift of the missing fraction to P_M as the material is solidified, and it too changes its rigidity and its ability for diffusion, without altering ordering or bonding. Missing fractions are dealt with separately in section 5; but it is likely that all the results given in Table 6.8 will have some bearing on the eventual emergence of a detailed and acceptable mechanism of muonium formation.

6.4 Mechanism of muonium formation

Hot model

Intertwined with the data presented in the two preceding sections has been a tacit acceptance that the muon's initial disbursement between diamagnetic and paramagnetic states was determined *during* the thermalization. In other words, it is the variety of epithermal interactions, including charge-exchange and hot-atom reactions of abstraction or substitution at the very ends of the track, which determines the muon's initial associations. This picture constitutes the rudiments of a so-called 'hot model' mechanism of muonium formation. It depends more on the detailed chemical nature of the molecules of the medium

Table 6.8. *Effect of the liquid–solid phase transition on the muon yields for a range of substances from highly-polar to non-polar*

Substance	Phase	T (K)	P_D	P_M	P_L
H_2O	liquid	295	0.62	0.20	0.18
	solid	270	0.48	0.52	0
NH_3	liquid	210	0.58	0.21	0.21
	solid	200	0.6	0.21	0.2
$C(CH_3)_4$	liquid	295	0.55	0.18	0.27
	solid	209	0.61	0.19	0.20
Xe	liquid	162	0.03	0.43	0.54
	solid	150	0.05	0.79	0.16

and less on its bulk physical properties. It is supported particularly clearly by the contrast amongst the noble gases in each phase (see Table 6.6 and Figure 6.6) and by the uniformity amongst the saturated hydrogenated liquids. It is possible, however, that the positive muon always emerges from this trail of primary events completely unassociated. If this happens, then it starts its chemical existence as a free μ^+ ion, neither as free Mu, nor already incorporated in diamagnetic molecules or free radicals. This situation is readily understandable for helium and neon because, as already discussed, reaction [6.2] dominates over [6.1] in charge exchange. No atom reactions are possible with He or Ne and μ^+ is virtually the only energetically-allowed and chemically-significant state. But this possibility has been postulated to occur in general, and has been specifically considered as a 'spur model' in water (Percival, Roduner & Fischer, 1978).

Spur model mechanism

Based on the effects of solute concentration and magnetic field on the observed P_D values in aqueous solutions, Percival *et al.* (1978, 1979; and Percival (1981*a*,*b*)) have proposed a spur model to account for the formation of muonium. This interesting notion was inspired by an analogous model for positronium formation (Mogensen, 1974). The idea is basically this: the μ^+ reaches thermal energies, unassociated chemically, but in close contact with the reactive species of the terminal spur of the muon track. It then starts to react with them before they disperse by diffusion, and if μ^+ combines with one of the free electrons then Mu will be formed. If μ^+ escapes such neutralization then it will not form Mu and, in general, will finally appear in a diamagnetic state.

The concept of radiation being deposited in 'packets', called spurs or δ-rays, is an old one (Lea, 1946) and stems from the fact that high-energy ionizing particles transfer energy by ionizing the medium and generating secondary electrons with a variety of kinetic energies. These secondary electrons, in turn, may cause further ionizations and molecular excitations. They tend to do so quickly because the range of an electron with a few hundred eV of energy is only a few nm. Spurs range in size from single ion-pairs to multi-ionizations (Mozumder & Magee, 1966). As a guideline, they are often regarded as averaging about five initial ion-pairs, corresponding to the deposition of $\sim 10^2$ eV, in a volume of ~1-2 nm radius (though obviously not often spherical). The closeness with which average spurs are formed down the track of the primary particle is represented by $-\mathrm{d}E/\mathrm{d}l$, the linear energy transfer (LET). It changes with the energy of the particle: increasing as the energy falls in accordance with the Bethe equation; passing through a maximum as the particle and orbital electron energies correspond, so that charge-exchange takes over; then falling sharply at the very end of the track (Mozumder, 1969). With water as the medium, the principal initial spur species are H_2O^+, e^- and H_2O^* in low-LET tracks. Within a picosecond or so, H_2O^+ should transfer a proton to give $OH + H_3O^+$, e^- gets localized then solvated, and H_2O^* can relax or dissociate. Some of the electrons interact with OH, H_3O^+ (or even H_2O^+) in which case they are lost and described as 'geminate'. The rest diffuse into the bulk of the medium and constitute the 'free ions'. The relative yields of free and geminate electrons varies enormously from one solvent to the next, as seen in Table 6.7, where these represent the average of all the spurs of a particle's track.

When considering the spur model of muonium formation one is focusing on the terminal spur and assuming μ^+ has stopped within it. These spurs will not all be of the same size, so again one is working with averages. Some will be single ion-pair spurs, others could be large. The spur model can be depicted by the scheme in Figure 6.9. Mu is formed by the intraspur reaction [6.6],

$$\mu^+ + e^- \rightarrow \text{Mu} \qquad [6.6]$$

which is in competition with the other electron reactions [6.7] and [6.8],

$$H_3O^+ + e^- \rightarrow [H_3O] \rightarrow H \qquad [6.7]$$

$$OH + e^- \rightarrow OH^- \qquad [6.8]$$

and with escape of μ^+ to give diamagnetic states through [6.9].

$$\mu^+ \xrightarrow{\text{aq}} \mu^+_{\text{aq}} \xrightarrow{H_2O} \text{MuOH} + H^+_{\text{aq}} \qquad [6.9]$$

Whether or not the electrons (and μ^+) of Eq. [6.6] are quasi-free, localized, weakly trapped, or fully solvated is a moot point (Walker, Jean & Fleming, 1980). Radiation chemical studies suggest that only ~15% on average of the geminate electron-decays precede solvation in low LET tracks (Jonah *et al.*, 1976); but Percival (1980 and 1981*a*) feel that reaction [6.6] precedes solvation by either species.

The evidence presented for the spur model comes from detailed analysis of the observed increase in P_D as a function of the applied transverse magnetic field and the concentration of scavengers [S] for both Mu and e^-. An example of typical data is given for chromate in Figure 6.10 (Percival, 1981*a*). Analogous curves were found for NO_3^-, MnO_4^- and oxalate, but the changes in P_D occurred in different concentration ranges for each solute. These data were analysed by 'fitting' to an equation of the form of [6.10],

$$P_D = h_D + h_M f(H, k[S]) \qquad [6.10]$$

in which $f(H, k[S])$ represents the field and solute concentration dependence of the observable conversion Mu → D prior to dephasing, h_D and h_M represent the post-spur fractional yields of D and Mu respectively. Best fits were obtained when h_D was allowed to increase. An increase in h_D (and corresponding decrease in h_M) is consistent with the solute inhibiting the intraspur formation of

Mu by reaction with the e^- of reaction [6.6]. Thus, the spur model's mechanism is basically that shown in Eq. [6.11],

$$\begin{array}{ccc} [e^- + \mu^+] & \xrightarrow{(h_M)} & Mu \\ k_e[S] \swarrow \quad \searrow (h_D) & & \swarrow k_M[S] \\ S^- \quad D & & \end{array} \qquad [6.11]$$

in which S reacts with e^- and Mu with rate constants k_e and k_M respectively. A solute which converts Mu to D but which is unreactive towards e^-, such as

Figure 6.9. (a) Representation of the spur model of Percival *et al.* (1978, 1979, 1981). (b) Mechanism showing interconversion of Mu→D and inhibition of Mu by electron scavenging in accordance with the spur model. (c) Simple interconversion (no inhibition) as in the mechanism of the hot model.

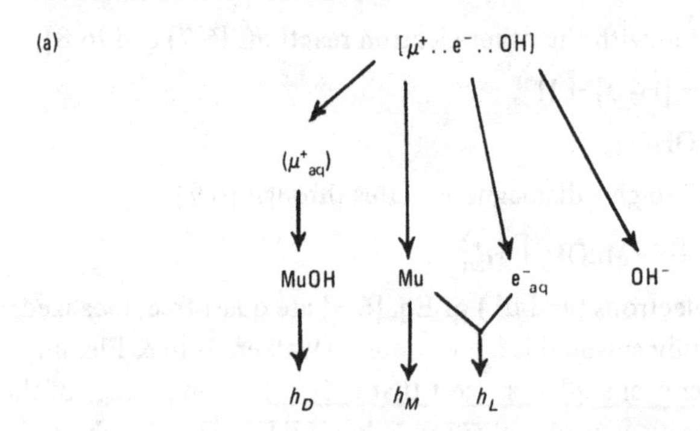

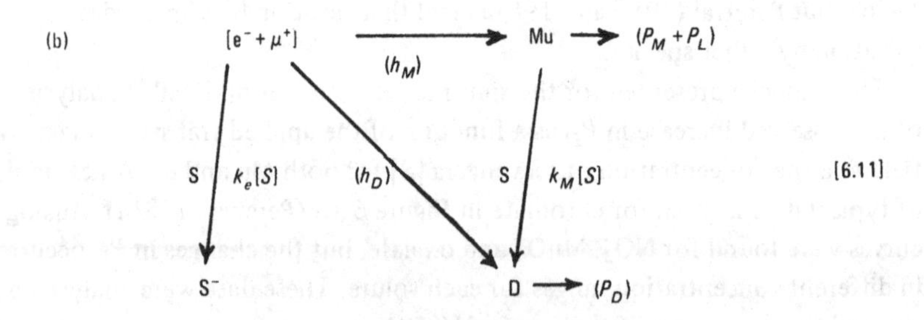

(c) μ^+ or Mu^* $\xrightarrow{(h_M)}$ Mu → $(P_M + P_L)$; (h_D); S $k_M[S]$; D → (P_D) [6.12]

$S_2O_3^{2-}$, increases P_D without altering h_D and h_M. The hot model (using h_D and h_M as *initial* yields) follows Eq. [6.12]

[6.12]

For solutes which react with both Mu and e^-, the values of h_D given in Figure 6.11 were obtained from the best fits to the model of Eq. [6.11]. One can see that h_D increases sharply over a relatively small concentration range for each solute. However, it is one of the idiosyncrasies of this model that the concentration at which each scavenger enhances h_D varies enormously, yet these are equally good electron scavengers. While k_e varies by less than a factor of two (Anbar *et al.*, 1975) the concentration at which h_D increases sharply varies by a factor of fifty.

Indeed, it is perhaps as well to point out at this stage that the spur model is not entirely self-consistent, even for the half-dozen solutes in aqueous solution to which it has been applied. For instance, the data from solutions of $S_2O_3^{2-}$ (not an electron scavenger) show remarkably similar field- and concentration-dependences to those of CrO_4^{2-} (a powerful electron scavenger), as indicated in

Figure 6.10. P_D as a function of transverse magnetic field H(G) at various concentrations of the good electron and Mu scavenger CrO_4^{2-} in water. Concentrations from top to bottom are: 1.0, 0.4, 0.2, 0.08, 0.04 and 0.01 M. (Figure taken from Percival, 1981*a*.)

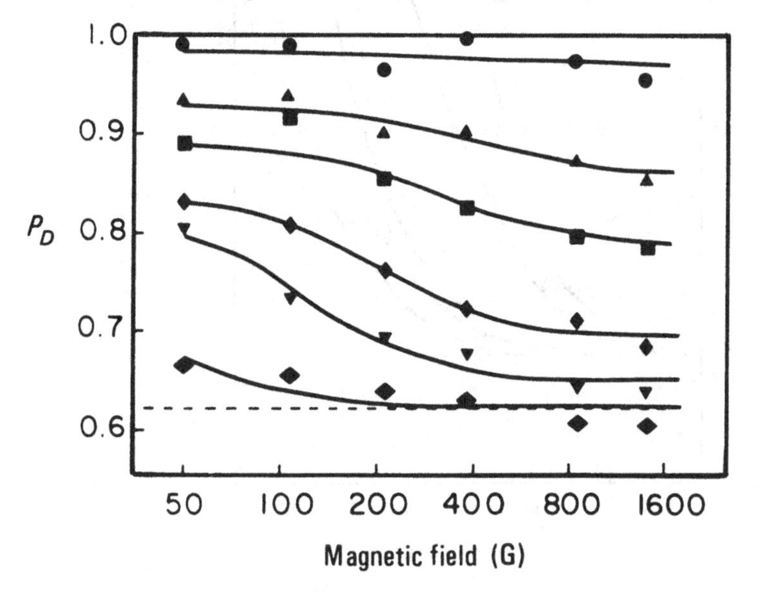

Figure 6.12 which is constructed from Figures 4.4 and 6.10. If the 'best fits' force these to have comparable changing values of h_D, then, evidently, they do *not* arise through inhibition of Mu formation by electron scavenging (Eq. [6.11]) because $S_2O_3^{2-}$ simply cannot intercept intraspur electrons significantly in the concentration range 0.1 to 1 M (as k_e is no more than 5×10^8 M^{-1} s^{-1}, Percival, 1981*b*; Buxton & Walker, unpublished data 1982).

Another inconsistency in the spur model is the fact that P_M is observed to be essentially unchanged by the presence of up to 1 M Cd^{2+} (Walker *et al.*, 1979). Of the various electron scavengers used in this study which are known to eliminate intraspur reactions of solvated and presolvated electrons, Cd^{2+} is the only one which does not react significantly with Mu at $\leqslant$1 M. It is the only electron scavenger in whose presence one can make a *direct* test of the spur model - by

Figure 6.11. Changes in the initial diamagnetic fraction h_D as determined by fitting P_D data to the mechanism of Figure 6.9 (b) and Eq. [6.11], as a function of concentration, for different electron-scavenging solutes in aqueous solution. For clarity, points very close to the pure water value (dashed line) have been omitted. Solutes which are not electron scavengers, such as $S_2O_3^{2-}$, would appear coincident with the dashed line (h_D invariant). (Curves taken from Percival, 1981.)

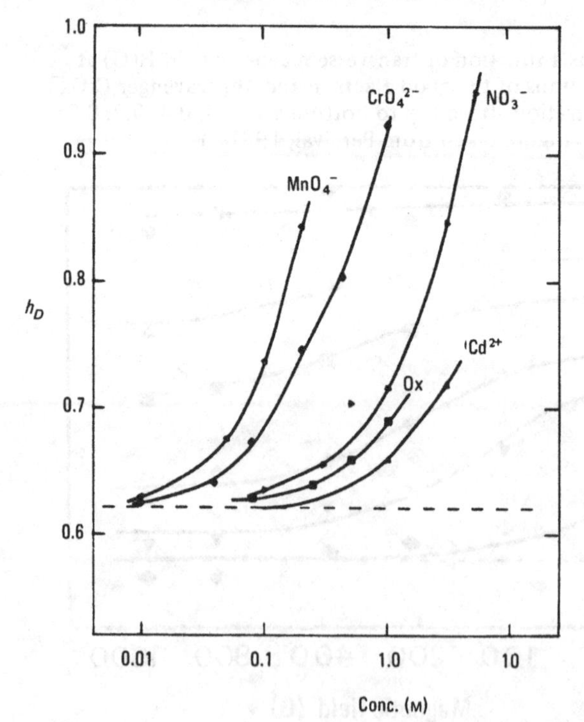

actually observing Mu – and when one does, P_M is found to be essentially unchanged. Cd^{2+} should be a particularly good discriminating solute for this purpose because the Cd^+ ion formed as a result of electron scavenging should not interact with μ^+, as it is known from picosecond-pulse radiolysis studies to react very slowly ($k = 10^5\,M^{-1}\,s^{-1}$, Wolff *et al.*, 1975) with H^+ and not to undergo secondary intraspur reactions (Hunt, 1976). It should also be noted that although Figure 6.11 shows h_D to increase by some 20%, this is only between 0.5 and 3.0 M $CdCl_2$ where the solutions are too concentrated (3 M $CdCl_2$ is ~50% by weight salt) to distinguish unambiguously 'spur' from 'hot'

Figure 6.12. Plot showing the lines of Figure 4.4 with the data points of Figure 6.10. The lines are the computer-fits for $S_2O_3^{2-}$ (which does not react with electrons so cannot have h_D changing) whereas the data points are for CrO_4^{2-} (which reacts with electrons and shows h_D increasing to 1). The lines (top to bottom) represent concentrations of 1.0, 0.5, 0.2, 0.1, 0.05, 0.02 and 0.01 M. (Data from Percival, 1981*b*); the data points (top to bottom, circle, triangle, square, diamond, inverted triangle and rotated square) are for 1.0, 0.4, 0.2, 0.08, 0.04 and 0.01 M, respectively. (Data from Percival, 1981*a*.) The uncertainty in the calibration of each P_D scale is ~± 0.03.

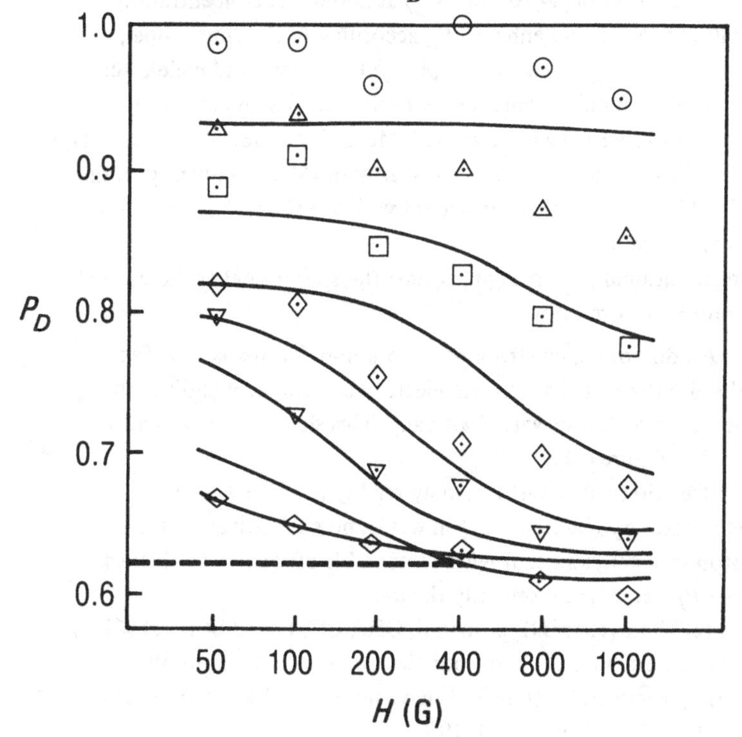

processes. Such high concentrations of efficient electron scavengers can equally be expected to interfere with epithermal processes, including the charge-exchange cycles, and thus to alter the primary yields within the context of the hot model. It is known, for instance, that it takes more Xe to form Mu in a He medium than in Ne - probably due to the different moderating efficiencies (Fleming *et al.*, 1979). Even the value of k in Eq. [6.10] may change with [S] as the solute probes earlier (hotter) stages of thermalization.

In addition, although equations such as [6.10] can take full account of the fact that the conversion reaction Mu → D is in competition with 'dephasing' for $h_M/2$ of the initial Mu ensemble (^{T}Mu), and with 'depolarization' through hyperfine oscillations for the other $h_M/2$(^{S}Mu); there is the unresolved matter (Percival *et al.*, 1978, 1979) of incorporating an initial 'missing fraction' into the analysis which should be consistent for all solutes.

It is when one puts the spur model to the test by examining additives which react with Mu *or* e^- but *not* both, that it seems to fail each time. Thus one has the following record: (a) For $S_2O_3^{2-}$, which reacts with Mu but not with e^-, one finds a similar field- and concentration-dependence as with scavengers of both - as discussed above; (b) For Cd^{2+} at $\leqslant 1$ M, which reacts with e^- but not with Mu, P_M and P_D remain virtually unchanged - also as described above; and (c), any powerful electron scavenger ($k_e \geqslant 10^{10}\,M^{-1}\,s^{-1}$) at about 1 M concentration should inhibit Mu formation and enhance P_D according to the spur model, regardless of how the scavenger reacts with Mu: yet Ni^{2+} ions and maleic acid, for instance, show no such enhancement (see Table 6.5). These two solutes happen to convert Mu to paramagnetic states (^{S}Mu and Mu-radicals, respectively) rather than to D; so their failure to change P_D, as with Cd^{2+}, is contrary to the spur model of Eq. [6.11] but entirely in accord with the simple mechanism of Eq. [6.12] and the hot model.

The following additional factors argue against the spur model as the general mechanism of muonium formation:

(i) P_M and P_D do not change from polar to non-polar media (see Table 6.7); despite the fact that ion and electron survival probabilities in radiation chemical spurs vary drastically. This sharp contrast is also presented in Figure 6.13.

(ii) Whereas free thermalized μ^+ ions may readily produce stable dia-magnetic states by reaction [6.9] in water, no such solvation-followed-by-proton-transfer can occur with comparable efficiency in alkanes. Yet their P_D values are essentially the same.

(iii) OH and H_3O^+ scavengers (2-propanol, OH^-, Cl^-, I^- and SCN^- at $\leqslant 1$ M) which should enhance the intraspur electron concentrations by eliminating reactions [6.7] or [6.8] have no observable effect on P_M (Walker, Jean & Fleming, 1979, 1980).

(iv) External electric fields (see Figure 6.8) which are known to alter the electron escape probability in radiation spurs (Allen, 1976*a*) and positronium formation (Mogensen, 1975), do not affect any of the measured P_D values (Ito *et al.*, 1981).

(v) Solvent mixtures show a nearly linear change in P_D (see Figure 6.5) even when just one of them is a powerful electron scavenger, like CCl_4 (Jean *et al.*, 1981).

(vi) P_D changes with the degree of halogenation in the series CH_2Cl_2, $CHCl_3$ and CCl_4 (see Figure 6.4) despite these three being equally good electron scavengers (Anbar *et al.*, 1975).

(vii) P_M does not change at the critical micelle concentration, indicating that $\mu^+ + e^-$ combination does not extend over distances comparable to those operative in $e^+ + e^-$ (see chapter 8, section 8).

Fig. 6.13. Plot showing the absence of any correspondence between either P_D and F (free ions), or P_M and F (geminate ions). F gives the fractional radiation yields, G/G(total). The dashed line **1** and **2** indicate, respectively, how P_D and P_M would have changed had there been a linear relationship between these muon yields and the corresponding electron escape probabilities from typical low *LET* spurs to which the F-values refer. Arrows indicate the following pure liquids: a is H_2O, b is CH_3OH, c is C_2H_5OH, d is $C(CH_3)_4$, e is $Si(CH_3)_4$, f is c-C_6H_{12} and n-C_6H_{14}.

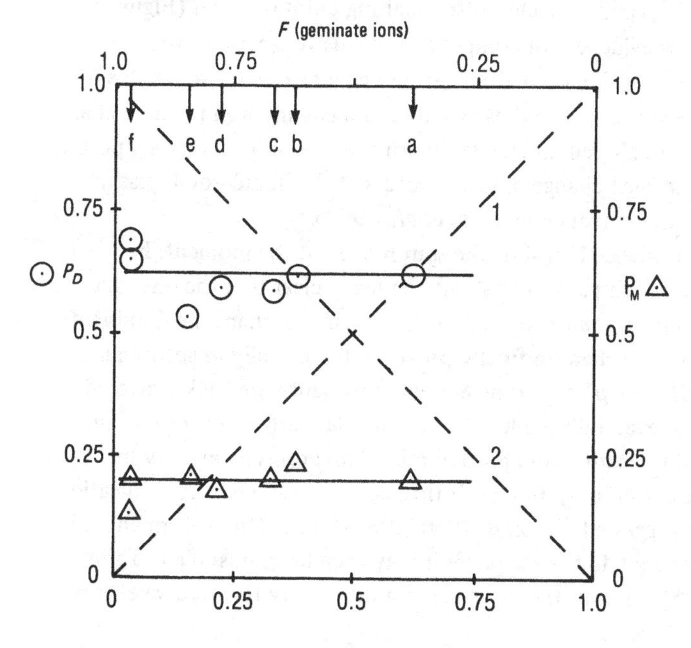

(viii) No anti-inhibition by C_6F_6 is found in *c*-hexane solutions with or without CCl_4 (Miyake *et al.*, unpublished data), in contrast to positronium.

There are some experimental results which seem to have ambiguous interpretations on this dichotomy of spur-versus-hot-models, including the following three: (a) P_D increases slightly with temperature in water (see Figure 6.7). An observed temperature-dependence usually suggests that a process is not governed by hot-atom reactions, on the grounds that a change in temperature of a hundred degrees or so (<0.1 eV) changes the energy of the 'hot' particle (often ≫1 eV) by a negligible amount. However, such arguments ignore the effect of the temperature-change on the medium. In the case of water the extent of structuring and hydrogen-bonding changes drastically from 0 to 100°C, and this could account for the observed temperature-dependence of P_D. (b) Most of those compounds in which Mu can be observed to have P_M values of ~0.2 in the liquid phase, seem to have P_M values of ~1.0 in the gas phase at ~1 atmosphere (101 kN m^{-2}) pressure (Arseneau *et al.*, 1982). Since radiation-chemical spurs are virtually non-existent in gases at such low pressures, it could be argued that the contrast in yields between liquid and gaseous phases reflects a contrast in formation mechanism. This would not be inconsistent with a spur model for liquids. However, there are other possibilities. One is the greatly reduced probability of hot-atom abstraction or substitution reactions by a particle as light as Mu* in the gas phase, with the result that they thermalize instead in that phase. (c) The sharp change in P_M, P_D and P_L yields at the melting point of water (Figure 6.7) may be interpreted as evidence for competitive, diffusive reactions, such as required by the spur model, with neutralization being favoured over escape in the change to the rigid phase. But these yields could equally well reflect a change in hot processes due to altered molecular ordering and hydrogen-bonding peculiar to water - because no such change in yields occurs at the liquid-solid (glassy) transition temperature in neopentane (Ito *et al.*, 1982).

On balance, the evidence is against the spur model, at the moment. But as a final comment on this comparison, it should be made clear that the discussion is concerned only about the *major* formation mechanism - perhaps the dominant 75(±15)% path. If the mechanism finally proves to be basically in accordance with the hot model, some μ^+ and some Mu will, by chance, be back-scattered into the muon track, even into the heart of a spur. Similarly, some hot-atom reactions are bound to accrue from excited muonium atoms formed by intraspur neutralization in reaction [6.6]. In fact, if this latter reaction precedes solvation by both species, as suggested (Percival, 1980, 1981*a*), then Mu* is formed with 13.5 eV of excess energy. In the gas phase it may even be impossible to form Mu by reaction [6.6], without the presence of a third body to conserve energy and momentum.

6.5 Comparison of the 'spur model' applied to muonium and positronium formation

If the spur model in Mu formation is to be rejected, then it is encumbent upon one to account for the differences between muonium and positronium formation - because in the latter system, the spur model enjoys considerable success (despite initial scepticism!), particularly on the effects of solutes in any given solvent (Maddock, Abbe & Haessler, 1977; Ito & Tabata, 1979). There are the following five points favouring a spur model as a major source of positronium, relative to muonium.

First, is the fact that the terminal spur of a positron's track will probably be created by the positron itself, whereas that in the muon's track will doubtless arise from a secondary electron, at least, if it is to be a multi-ionization spur. An attempt to compare typical end-of-track events is presented in Figure 6.14. If this is at all realistic, then it is easy to see why e^+ can often finish up within the terminal spur and why μ^+ will generally not do so.

Second, during the last $\sim 10^3$ eV of energy, the track of e^+ is expected to be of relatively high LET, while that of μ^+ is of low LET. This is shown in the plots presented by Mozumder (1969) and reproduced in Figure 6.15. It means that there is a much denser track of charged species to be attracted or scattered back into, in the wake of the positron than of the muon. Indeed, there is a non-negligible chance that e^+ will devote its last 10^2 eV to forming a spur, as in Figure 6.14. But that is highly unlikely for μ^+ in view of charge-exchange, hot-atom processes, and in the spur formation mechanism suggested in Figure 6.14.

Third, the possibility of forming hot Ps by charge-exchange cycles during the slowing down of e^+ is expected to be totally different from that of Mu. If one made a mass extrapolation on Figures 6.1 and 6.2 then charge-exchange would commence only at ~ 50 eV, rather than at $\sim 10^4$ eV for μ^+ (though such an extrapolation is not really valid because the Born approximation obviously cannot be used). In fact, Ps formation is believed to occur mainly in the narrow Ore gap (Jean & Ache, 1979) given by Eq. [6.13]

$$I(X) < E_k < I(X) - I(Ps) \quad [6.13]$$

where E_k is the kinetic energy of e^+, $I(X)$ is the ionization energy of the molecules of the medium (generally, 8-13 eV) and $I(Ps)$ the ionization energy of Ps (6.8 eV). For the range $E_k > I(X)$, ionization of X is still the preferred process (Jean & Ache, 1979); whereas for the last few eV of its track, the conversion from $e^+ \rightarrow Ps$ is endoergic and therefore improbable. The scale below the abscissa in Figure 6.1 shows the limited energy range over which Ps may form according to Eq. [6.13]. By contrast, Mu formation from μ^+ commences at much higher energy, and remains exoergic until it reaches thermal energy -

except in He and Ne, as discussed earlier. Indeed, in these two systems charge-exchange is also seen to be a one-way process at the end (see Figures 6.1 and 6.2) since all the muons emerge as bare μ^+ (or are associated weakly in diamagnetic species such as $NeMu^+$ (Mikula, 1981).)

Fourth, hot atom reactions of Ps* are much less probable than for Mu*: firstly, because hot Ps exists only over a very much more restricted energy range (I(X)–6.8 eV) as compared to 10^4 eV; secondly, hot-atom reactions of abstraction and substitution by Ps* should occur with even lower efficiency than Mu* because of the former's extremely low mass compared to the mass of the atom being abstracted or replaced.

Figure 6.14. Comparison of plausible last spurs in the tracks of (a) a positron, and (b) a muon. As indicated in (a), the positron *itself* probably creates its last spur (outlined by the dashed circle) and therefore may very well finish up within that spur. By contrast, the last spur in the muon track is created by a secondary electron (δ-ray) of sufficient energy. Being much more massive, the muon ploughs on causing further ionizations and excitation, undergoing charge-exchange and eventually thermalizing some distance beyond the last spur as μ^+ or Mu.

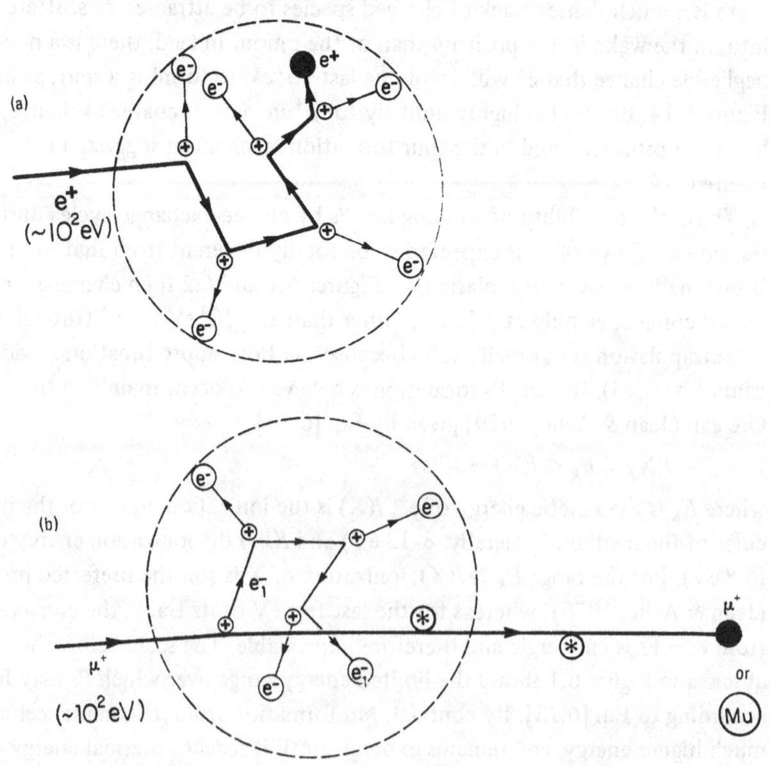

Finally, the mobility of the quasi-free positron has recently been found to be exceptionally large in hydrocarbons (Heinrich & Schiltz, 1982), higher even than quasi-free electrons in spherical non-polar molecules. This makes e^+ particularly susceptible to capture by e^- at the very end of the track, even if it were thermalized some distance from the terminal spur. However, the mobility of μ^+ should be smaller than that of e^+ by at least the square root of the mass ratio. Furthermore, e^+ being much lighter than μ^+, is more readily backscattered by the medium, influenced by the net attractive coulomb field of the ionization track, and has a more extended de Broglie wavelength for tunnelling and relocation in its pre-localized state.

For the above reasons the end of track events of e^+ and μ^+ are so different that it seems perfectly acceptable to have a spur model applying to one but not the other.

6.6 Missing fractions

As soon as muonium was first detected directly in water it was realized that there was a 'missing fraction' (P_L) of muon polarization (Percival *et al.*, 1976). There are reasons for believing that this does not stem from depolariza-

Figure 6.15. Variation of *LET* (linear energy transfer in eV/nm, or 100 eV spurs per 100 nm) with particle energy, *E* (in eV), shown as a log–log plot for water as the stopping medium. Curve **1** refers to the positive muon, and curve **2** to the electron. N.B. At about 300 eV, the *LET* of μ^+ is <1 eV/nm, or one 100-eV spur every 100 nm, whereas for e^- (e^+ expected to be slightly larger) the *LET* is ~ 70 eV/nm, or one 100-eV spur every 1.5 nm. (Curves taken from Mozumder, 1971.)

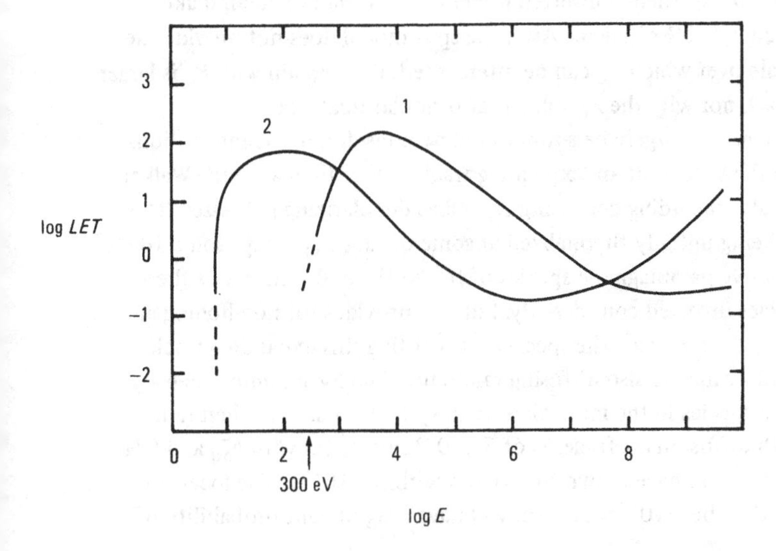

tion through hyperfine oscillations in the ^{S}Mu state formed during the charge-exchange cycles of the thermalization process (Massey & Burop, 1969). Thermalization occurs too quickly in the condensed phase for this to contribute significantly, and, in any case, values of P_L range from zero in CCl_4 to 0.8 in CS_2. So one has to look elsewhere. In some systems P_L doubtless arises from chemical reactions of the solvent or its impurities with Mu or Mu-radicals. But there are several media, such as the saturated hydrogenated compounds in the Appendix, where Mu is observed on the μSR time scale, and for these liquids P_L and P_M are approximately equal.

A missing fraction can genuinely arise from chemical interactions occurring on the timescale $<10^{-7}$ s which produce either: (i) a dephasing of the muon spin by interconversion of muon states having different precession periods; or (ii) depolarization of the muon spin by a Heisenberg spin-exchange interaction. It is fairly certain too that the missing fraction in water was part of the initial muonium fraction, because additives which rapidly convert Mu → D regenerate all the muon polarization by increasing P_D to 1.0 (see Figures 4.4 and 6.10, and Table 6.5).

Percival *et al.* (1978, 1979) included an explanation for the missing fraction in their spur model, in the form of a depolarizing second encounter of Mu with a paramagnetic species (principally e_s^-). This proposal has the great merit that it provides for interactions of very limited duration, as necessitated by the fact that Mu is seen later ($\sim 10^{-6}$ s). However, it does not solve the problem of how the muon invariably found itself within a sufficiently large spur to experience two consecutive encounters with high probability. As half the initial Mu ensemble seems to be lost, then either they all make a potentially-depolarizing encounter with 50% of them converted to ^{S}Mu, or else half of them make an encounter which is 100% efficient. Also, the spur model does not provide the correct timescale over which P_L can be intercepted, nor explain why P_L is larger in D_2O than H_2O, nor why the P_L value is also similar in alkanes.

These problems can largely be surmounted by considering the interactions occurring with the species of an 'expanding track' rather than a 'spur' (Walker, 1981); and by also including dephasing as well as depolarizing processes. This allows for Mu being initially thermalized at some distance (perhaps some 10 nm) beyond the reactive paramagnetic species of the track. It also increases the number of species involved considerably, but still provides for non-homogeneous kinetics. As they diffuse apart the species constituting this expanding track overlap with Mu – which is also diffusing randomly. The local number density of paramagnetic species in the immediate vicinity of Mu may rise then fall. For species with diffusion coefficients of 5×10^{-5} cm^2 s^{-1} (as with e_{aq}^- and OH in water) the track will have grown to ~10 nm within ~10^{-8} s. The local concentration will then be ~10^{-3} M and Mu will have a significant probability of an encounter within 10^{-7}s, but trivial thereafter. This is depicted in the sketch of Figure 6.16. It means that solutes which scavenge the reactive species – which in the case of the alkanes are mainly atoms and radicals – will eliminate P_L while enhancing P_M, except when the product of reaction may also depolarize Mu or convert Mu → D.

An essential difference between the 'intraspur' and 'expanding track' pictures is that in the former Mu is in competition with geminate processes, whereas in the latter, Mu encounters species which *escape* from the spurs. Therefore, only the expanding track accounts for the greater missing fraction in D_2O than in H_2O (Table 6.6), because the free-ion yield is larger in D_2O (due to its slower dielectric relaxation). Also, the timescale is more appropriate for the changes observed at freezing points. Thus, solidification can readily alter the overlap and interaction time of an expanding track from 10^{-8}–10^{-7}s in the liquid to $\gg 10^{-5}$s in the solid, merely by reducing the diffusion coefficient from $\sim 10^{-4}$ to $<10^{-7}$ $cm^2 s^{-1}$. A change of phase to the solid, even in water with its prowess for crystallizing, can hardly be expected to 'freeze' intraspur processes. In fact,

Figure 6.16. Expanding track picture of Mu depolarization (or reaction). (a) The initial track is represented by 10 (corresponding to time of 10^{-10}s). The species of the track then expand approximately to the cylindrical volumes represented by lines 9, 8 and 7 at times of 10^{-9}, 10^{-8} and 10^{-7}s, respectively. Mu is initially thermalized at M (15 nm from the end of the track, as sketched here) then it diffuses randomly. Averaged over the ensemble it will generally be found in spherical volumes bounded by the lines 9, 8 and 7 at 10^{-9}, 10^{-8} and 10^{-7}s later. (b) Shows the variation of concentration ($\log c$) as a function of time ($-\log t$) for distances (d) of initial Mu localization of 0, 10, 20 and 30 nm from the end of the track. No account is taken of intraspur reactions in estimating c for Fig. 6.16(b).

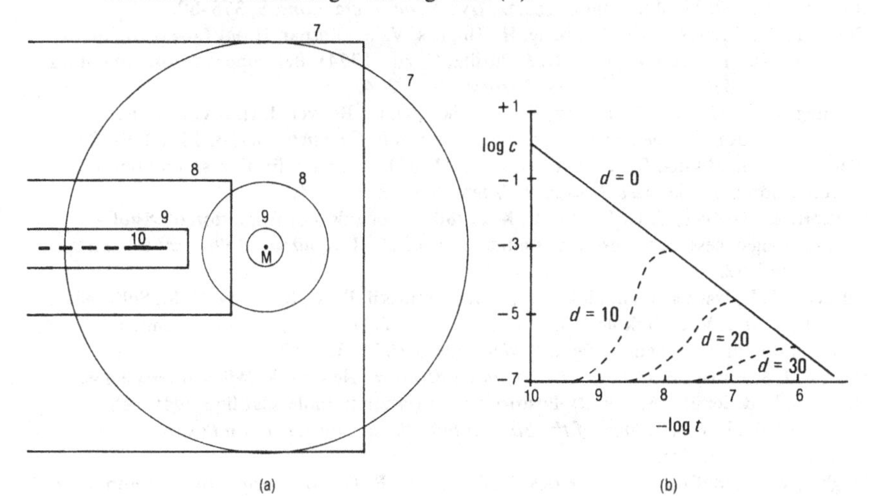

intraspur electron recombination or escape processes have been observed to be complete in single crystals of ice in $\leqslant 10^{-6}$ s (Gillis, Teather & Ross, 1980). Finally, the tentative evidence that at 85 K in liquid Ar one actually observes half of the Mu ensemble decay with a lifetime $\sim 1 \times 10^{-7}$ s (Keifl, Warren, Marshall & Oram, 1981) is inconsistent with the timescale of intraspur reactions, but not inconsistent with the period of expanding track overlap (see Figure 6.16).

References for chapter 6

Allen, A. O. (1976*a*). Yields of free ions formed in liquids by radiation. *National Standard Reference Data Systems - National Bureau of Standards*, No. 57.

Allen, A. O. (1976*b*). Drift mobilities and conduction band energies of excess electrons in dielectric liquids. *National Standard Reference Data Systems - National Bureau of Standards*, No. 58.

Anbar, M., Farhataziz & Ross, A. B. (1975). Selected specific rates of reactions of transients from water in aqueous solution. I. Hydrated electrons. *National Standard Reference Data Systems - National Bureau of Standards*, No. 43.

Arseneau, D. *et al.* (1982). Unpublished data.

Babaev, A. I., Balats, M. Ya., Myasishcheva, G. G., Obukhov, Yu. V., Roganov, V. S. & Firsov, V. G. (1966). An experimental investigation of chemical reactions involving muonium. *Soviet Physics JETP*, **23**, 583–91.

Barnett, B. A., Chang, C. Y., Yodh, G. B., Carroll, J. B., Eckhause, M., Heieh, C. S., Kane, J. R. & Spence, C. B. (1975). Muonium-formation measurement in low pressure argon gas. *Physical Review*, **A11**, 39–41.

Bell, R. P. (1973). *The Proton in Chemistry*, edition 2. Ithaca: Cornell University Press.

Brewer, J. H., Crowe, K. M., Gygax, F. N. & Schenck, A. (1975). Positive muons and muonium in matter. In *Muon Physics*, vol. 3, ed. V. W. Hughes & C. S. Wu, pp. 3–139. New York: Academic Press.

Brewer, J. H. & Crowe, K. M. (1978). Advances in muon spin rotation. *Annual Reviews of Nuclear and Particle Science*, **28**, 239–326.

Brewer, J. H. (1981). Muonium in quartz. *Hyperfine Interactions*, **8**, 375–80.

Crane, T. W., Casperson, D. E., Chang, H., Hughes, V. W., Kaspar, H. F., Lovett, B., Schiz, A., Sonder, P., Stambaugh, R. D. & Putlitz, G. zu. (1974). Behaviour of positive muons in liquid helium. *Physical Review Letters*, **33**, 572–4.

Fleming, D. G., Garner, D. M., Vaz, L. C., Walker, D. C., Brewer, J. H. & Crowe, K. M. (1979). Muonium chemistry - a review. *Advances in Chemistry Series*, **175**, 279–334.

Fleming, D. G., Mikula, R. J. & Garner, D. M. (1981). μ^+ thermalization and muonium formation in noble gases. *Hyperfine Interactions*, **8**, 307–14.

Gillis, H. A., Teather, G. G. & Ross, C. K. (1980). Mechanism of formation of visible-absorbing excess electrons in crystalline ice near 273 K. *Journal of Physical Chemistry*, **84**, 1248–52.

Gurevich, I. I., Ivanter, I. G., Meleshko, E. A., Nikolskii, B. A., Roganov, V. S., Selivanov, V. I., Smilga, V. P., Sokolov, B. V. & Shestakov, V. D. (1971). Two-frequency precession of muonium in a magnetic field. *Soviet Physics JETP*, **33**, 253–9.

Hart, E. J. & Anbar, M. (1970). *The Hydrated Electron*. New York: Wiley-Interscience.

Heinrich, F. & Schiltz, A. (1982). Positron implantation in molecular liquids at high electric fields. *Proceedings of the Sixth International Conference on Positron Annihilation*, Arlington, Texas.

Hughes, V. W., McColm, D. W., Ziock, K. & Prepost, R. (1960). Formation of muonium and observation of its Larmor precession. *Physical Review Letters*, **5**, 63–5.

Hunt, J. W. (1976). Early events in radiation chemistry. In *Advances in Radiation Chemistry* vol. **5**, ed. M. Burton & J. L. Magee, pp. 185–315. New York: Wiley-Interscience.
Ito, Y. & Tabata, Y. (1979). Positronium formation experiments in organic liquids. Effects of electron scavengers. *Proceedings of 5th International Conference on Positron Annihilation*, Lake Yamanake, Japan, pp. 321–4.
Ito, Y., Ng, B. W., Jean, Y. C. & Walker, D. C. (1980). Muonium atoms observed in liquid hydrocarbons. *Canadian Journal of Chemistry*, **58**, 2395–401.
Ito, Y., Ng, B. W., Jean, Y. C. & Walker, D. C. (1981). Effect of external electric fields on the μSR of liquid hydrocarbons and fused quartz. *Hyperfine Interactions*, **8**, 355–8.
Jean, Y. C. & Ache, H. J. (1979). Search for selectivity between optical isomers in the interactions of positrons with chiral molecules. In *Origins of Optical Activity in Nature*, ed. D. C. Walker, pp. 67–86. Amsterdam: Elsevier Scientific Publishing Company.
Jean, Y. C., Ng, B. W., Brewer, J. H., Fleming, D. G. & Walker, D. C. (1981). Origin of the residual polarization in muon chemistry studies of solvent mixtures. *Journal of Physical Chemistry*, **85**, 451–4.
Jean, Y. C. *et al.* (1982). Unpublished data.
Jonah, C. D., Matheson, M. S., Miller, J. R. & Hart, E. J. (1976). Yield and decay of the hydrated electron from 100 ps to 30 ns. *Journal of Physical Chemistry*, **80**, 1267–80.
Keifl, R. F., Warren, J. B., Marshall, G. M. & Oram, C. J. (1981). Muonium in the condensed phases of Ar, Kr, and Xe. *Journal of Chemical Physics*, **74**, 308–13.
Lea, D. E. (1946). *Actions of Radiations on Living Cells*. Cambridge: Cambridge University Press.
Maddock, A. G., Abbe, J. Ch. & Haessler, A. (1977). The chemistry of positronium. Inhibition and reactivity of solutes with electrons. *Chemical Physics Letters*, **47**, 314–18.
Manion, J. P. & Burton, M. (1952). Radiolysis of hydrocarbon mixtures. *Journal of Physical Chemistry*, **56**, 560–9.
Massey, H. S. W. & Burop, E. H. S. (1969). *Electronic and Ionic Impact Phenomena*. London: Oxford University Press.
Mikula, R. J. (1981). *Relaxation and Formation Processes of the Muon and Muonium in the Gas Phase*, Ph.D. Thesis, University of British Columbia, pp. 1–227.
Miyake, Y., Tabata, Y., Ito, Y. *et al.* (1982). (To be published.)
Mogensen, O. E. (1974). Spur reaction model of positronium formation. *Journal of Chemical Physics*, **60**, 998–1004.
Mogensen, O. E. (1975). Effect of an external electric field on the positronium formation in the positron spur. *Applied Physics*, **6**, 315–22.
Mozumder, A. & Magee, J. L. (1966). Model of tracks of ionizing radiations for radical reaction mechanisms. *Radiation Research*, **28**, 203–14.
Mozumder, A. (1969). Charged particle tracks and their structure. In *Advances in Radiation Chemistry*, vol. **1**, ed. M. Burton & J. L. Magee, pp. 1–102. New York: Wiley-Interscience.
Myasishcheva, G. G., Obukhov, Yu. V., Roganov, V. S. & Firsov, V. G. (1967). The chemistry of muonium, I and II. *High Energy Chemistry*, **1**, 389–93.
Ng, B. W. (1980). *Studies of the Muonium Atom in Liquid Media*, M.Sc. Thesis, University of British Columbia.
Ng, B. W., Stadlbauer, J. M., Jean, Y.-C. & Walker, D. C. (1982). Muonium atoms in liquid and solid neopentane. *Canadian Journal of Chemistry* (in press).
Percival, P. W., Fischer, H., Camani, M., Gygax, F. N., Ruegg, W., Schenck, A., Schilling, H. & Graf, H. (1976). The detection of muonium in water. *Chemical Physics Letters*, **39**, 333–5.
Percival, P. W., Roduner, E. & Fischer, H. (1978). Radiolysis effects in muonium chemistry. *Chemical Physics*, **32**, 353–67.
Percival, P. W., Roduner, E. & Fischer, H. (1979). Radiation chemistry and reaction kinetics of muonium in liquids. *Advances in Chemistry Series*, **175**, 335–55.
Percival, P. W. (1980). The time-scale of intraspur muonium formation. *Journal of Chemical Physics*, **72**, 2901–2.

Percival, P. W. (1981*a*). Muonium formation in water and aqueous solutions. *Hyperfine Interactions*, 8, 315–23.

Percival, P. W. (1981*b*). (i) Use of spin polarized muons to probe hydrogen atom reactions. *15th International Free Radical Symposium, Nova Scotia*. (ii) *Annual Report of TRIUMF (1981)*, University of B.C., Vancouver, 43–5.

Roduner, E., Percival, P. W., Fleming, D. G., Hochmann, J. & Fischer, H. (1978). Muonium-substituted transient radicals observed by muon spin rotation. *Chemical Physics Letters*, 57, 37–40.

Spinks, J. W. T. & Woods, R. J. (1976). *An Introduction to Radiation Chemistry*, 2nd edn, pp. 380–440. New York: Wiley-Interscience.

Stadlbauer, J. M. *et al.* (1982). Unpublished data.

Stambaugh, R. D., Casperson, D. E., Crane, T. W., Hughes, V. W., Kaspar, H. F., Sonder, P., Thompson, P. A., Orth, H., Putlitz, G. zu. & Denison, A. B. (1974). Muonium formation in noble gases and noble-gas mixtures. *Physical Review Letters*, 33, 568–71.

Swanson, R. A. (1958). Depolarization of positive muons in condensed matter. *Physical Review*, 112, 580–6.

Walker, D. C. (1981). Arguments against a spur model for muonium formation. *Hyperfine Interactions*, 8, 329–36.

Walker, D. C., Jean, Y. C. & Fleming, D. G. (1979). Muonium atoms and intraspur processes in water. *Journal of Chemical Physics*, 70, 4534–41.

Walker, D. C., Jean, Y. C. & Fleming, D. G. (1980). Reply to: the time-scale of intraspur muonium formation. *Journal of Chemical Physics*, 72, 2902–4.

Wolff, R. K., Aldrich, J. E., Penner, T. L. & Hunt, J. W. (1975). Picosecond pulse radiolysis. V. Yield of electrons in irradiated aqueous solution with high concentrations of scavengers. *Journal of Physical Chemistry*, 79, 210–19.

7

MUONIUM REACTIONS IN GASES

7.1 Preamble

As noted in the Introduction, muonium is ideal as a light isotope of hydrogen, and hydrogen is renowned for incredible versatility in its types of reactions. It is not unreasonable then, to hope that the unique sensitivity provided by the nine-fold mass ratio of H to Mu will unearth various mass-dependencies buried in kinetic parameters. It seems appropriate to start the study of muonium reactivity and kinetic isotope effects with gas phase reactions – particularly with those elementary reactions in which a free atom attacks a diatomic molecule at low pressure. This is where theory and experiment are closest. In fact, it is specifically for the case of muonium that theoretical descriptions of the energetics and dynamics of a chemical reaction are seen to have to undergo a major transition from purely classical to largely quantum mechanical in nature, at room temperature.

Kinetic isotope effects will be reported as k_M/k_H, the ratio of rate constants for the reaction of Mu relative to ^1H with a given substrate. There are several reasons why this ratio should differ from unity. First, is the difference in the mean velocity of Mu compared with H. Given that the mean kinetic energies ($m\bar{v}^2/2$) will just be proportional to the temperature under the conditions of a μSR measurement, the mean velocity ($\bar{v}$) of Mu will exceed that of H by a factor of 3.0. In evaluating the encounter frequency with the substrate molecule, one requires the reacting partners' relative velocities. Therefore, it is the square root of the reduced mass ratio that is appropriate. This means that the encounter frequency of Mu compared to H is typically greater by a factor of ~2.9. (For the reaction with H_2 it is only 2.5.) Second, the collision will be of shorter duration by this factor of three, in the sense of the actual time spent by the atom within the interaction distance of the substrate. This can cause a reduced isotope effect for those reactions involving orientational factors when a specific rotational or vibrational mode of the substrate is required. Third, the smaller mass of Mu gives it greater zero-point motion. This confers a higher energy on the activated complex and on the reaction products, for Mu compared to H. Thus Mu reactions have the higher activation barriers, which tends to reduce the net isotope effect. Fourth, quantum mechanical tunnelling strongly favours Mu over H because the tunnelling probability has an inverse exponential reduced-mass dependence.

In the μSR method, the fate of ^{T}Mu is studied one-atom-at-a-time in the presence of a constant and relatively large concentration of substrate, $[S]$. A decay constant (λ) is evaluated from the accumulated lifetime histogram by fitting to an exponential decay in the number of ^{T}Mu atoms remaining as a function of time. These are ideal pseudo-first-order kinetic conditions: one is making a direct measurement of the rate of disappearance of the minor reactant. There is invariably a small 'background' decay constant (λ_0) in the absence of solute, as discussed in chapter 4, so the bimolecular rate constant for reaction of ^{T}Mu with S (k_M) is evaluated by use of Eq. [7.1].

$$k_M = (\lambda - \lambda_0)/[S] \qquad [7.1]$$

If λ_0 in fact arises from an experimental artifact such as an impurity, or magnetic field inhomogeneity or fluctuation, or from the cell itself, so that it remains constant, then Eq. [7.1] should be perfectly valid. But if it is a fitting problem arising from two-frequency beating, then it could change slightly with λ and a small additional error in k_M would be introduced. But since λ_0 is usually of the order of 10% of λ, the possible error from this source is much smaller than other random errors, and from typical statistical errors from the fitting procedure.

7.2 Rate constant measurements

Experimental aspects

Hughes (1966) and collaborators (Mobley *et al.*, 1966, 1967*a*; Mobley, 1967*b*), initiated the chemical study of Mu as an isotope of H following their discovery of muonium formation (Hughes, McColm, Ziock & Prepost, 1960). They used argon gas as the inert medium in which to produce Mu and to moderate it to the ambient thermal energy. Rates of reaction of ^{T}Mu were then observed in the presence of added gases, present in sufficiently small concentration to not affect Mu-formation nor react with it epithermally. These early experiments used high-energy muons ('backward' mode – see chapter 3) of such penetration that gas pressures up to 40 atmospheres (4 MN m^{-2}) were required to ensure that the muons stopped in the gas sample. Such high pressures permitted a very large range of substrate pressures to be studied, but they occasionally introduced the possibility of termolecular (third-body) effects. These authors used all three types of μSR measurement – rotation in transverse fields, relaxation in longitudinal fields and resonance in radiofrequency fields (see Figure 3.1).

Some of the problems arising from the need to use high pressures were averted by adopting the 4.1-MeV muon beam ('surface' mode, Pifer, Bowen & Kendal, 1976) in order to study gases at total pressures of about one atmosphere (Fleming *et al.*, 1976, 1979; Fleming, Mikula & Garner, 1980; Fleming, Garner & Mikula, 1981; Garner, Fleming & Brewer, 1978; Garner, 1979). Figure 7.1 shows a typical experimental arrangement. The volume of sample is still comparable in size to the high-pressure work, and its wide stopping region causes somewhat reduced asymmetry and a spread of initial phases compared to studies in the condensed phases. The improved technology gives these low-pressure studies higher counting statistics and broader scope. Several reactions have been studied as a function of temperature in order to obtain Arrhenius parameters. Both Ar and N_2 have been used as the formation and moderating medium for Mu, the latter giving a higher yield.

Results and their interpretation

Mobley *et al.* (1966, 1967*a*,*b*) surveyed most of the various muonium reaction types. They studied abstraction or substitution reactions with Cl_2, CH_3Cl and C_2H_6, and addition with C_2H_4. They also studied the reaction with the paramagnetic gases O_2 and NO. In principle O_2 and NO may combine

Figure 7.1. Diagram of the apparatus used to make μSR measurements in argon at about atmospheric pressure with 'surface' muon beams. The incident μ^+ triggers beam counters B2 and B3; later, e^+ may trigger either left (L1, L2 and L3) or right (R1, R2 and R3) positron telescopes. Two-inch graphite absorbers were used to stop low energy e^+. The magnetic field is vertically out of the paper. (Figure taken from Fleming *et al.*, 1976.)

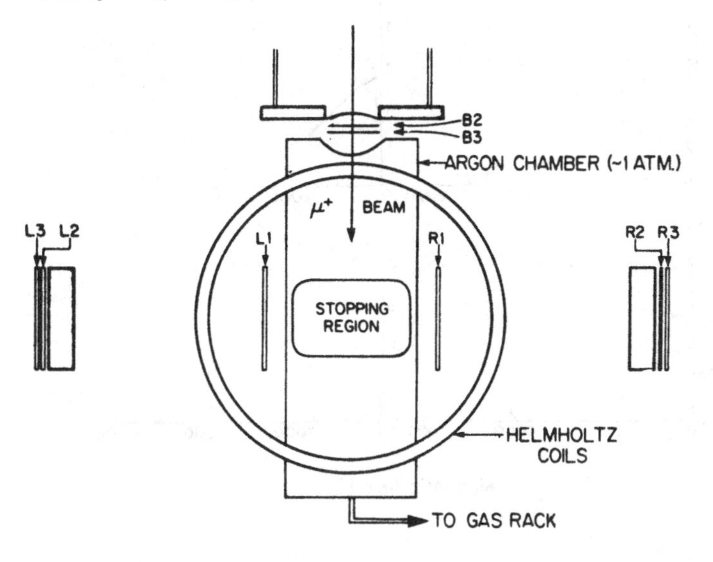

chemically with Mu; but Mobley *et al.* (1967*a,b*) deduced that the interaction was an electron spin-exchange, converting $^{T}Mu \rightarrow {}^{S}Mu$. Their conclusion was based on the field-dependence of the polarization in longitudinal fields, as shown in Figure 7.2, and its contrast with the reactions with diamagnetic molecules C_2H_4 and Cl_2. This assignment has recently been confirmed by excluding the possibility of termolecular chemical reactions through the use of low pressures (Fleming *et al.*, 1980; Mikula, Garner & Fleming, 1981). The absence of any activation energy or isotope effect other than the encounter frequency is also reported to be consistent with spin-exchange. It is not known yet whether these are actually electron spin-exchanges rather than simply catalysed spin-flips.

This type of Heisenberg outer-shell electron spin-exchange interaction is of growing interest (Molin, Salikhov & Zamaraev, 1980). It affords the theorist a system of intriguing simplicity, and it provides the experimentalist with a handle on certain complex types of processes in solution and in biological systems involving changes in spin state. Such changes can only rarely be studied by conventional methods. It is the unique spin characteristics of the μSR technique that allows this type of interaction to be followed so easily for muonium.

Recognizing its great interest from a theoretical point of view, and for com-

Figure 7.2. Observed values of λ/n (proportional to k_M) versus magnetic field for NO and O_2 reacting with Mu as studied in longitudinal magnetic field. The solid lines are theoretical relationships for a spin-exchange interaction ($k_M \propto [1-(H/1585\ G)^2]^{-1}$). For comparison purposes the data for the substrates C_2H_4 and NO_2 are given. (Figures taken from Mobley *et al.*, 1967*a*.)

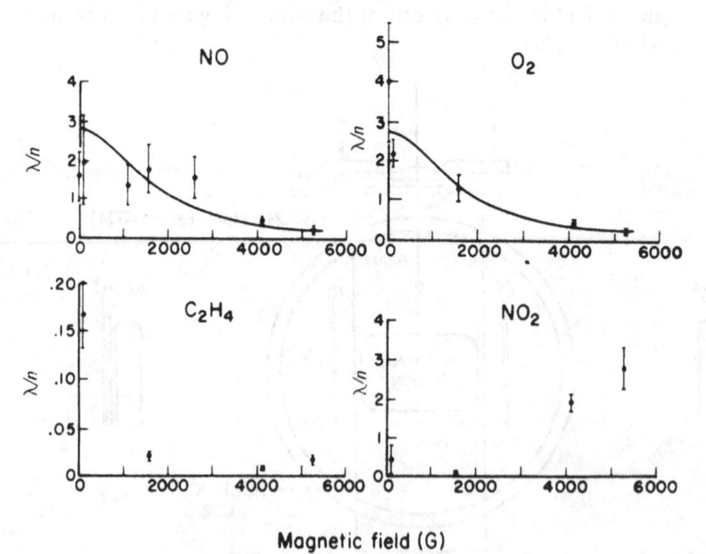

parison with its isotopic counterparts, Mobley *et al.* (1967*a*,*b*) studied the reaction of Mu with H_2, as in Eq. [7.2],

$$Mu + H_2 \rightarrow MuH + H \qquad [7.2]$$

up to high H_2 pressures; but they found the rate constant to be too small to measure at room temperature. However, using temperatures up to 850 K and pressures up to 5 atmospheres (0.5 MN m^{-2}), Garner *et al.* (1982) have recently determined the main kinetic parameters for this reaction and its D_2 analogue. Their preliminary results are given in Figure 7.3, where they are also compared with the variational transition-state theoretical treatment of Bondi *et al.* (1982). Good agreement is found (see later). Whereas the $H + H_2$ reaction is thermoneutral, that of $Mu + H_2$ is endothermic (by ~25 kJ mol^{-1}) due to the higher zero-point energy in MuH compared to H_2. This results in a much larger activation barrier, with the net result than an inverse isotope effect of 0.055 is found for reaction [7.2] at 800 K. Here is a fine example of the zero-point energy effect strongly favouring the heavier isotope, because it is done under sufficiently high temperature conditions for tunnelling to be of minor importance.

Figure 7.3. Experimental and theoretical rate constants for the reactions $H + H_2$ (top), $Mu + H_2$ (middle) and $Mu + D_2$ (bottom). The solid lines connect the experimental data points (Mu data points from Garner *et al.*, 1982); the dots represent the calculations of Bondi *et al.* (1982). (Data taken from Fleming, 1981.)

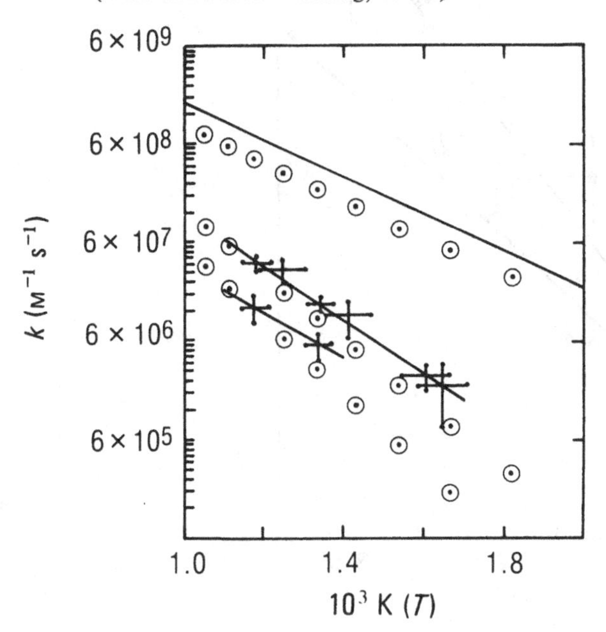

Reactions of Mu with the halogens are also of particular interest, showing as they do, other isotope effects. An investigation of the reaction of Mu with the halogens F_2, Cl_2 and Br_2, and the halogen acids HCl, HBr and HI, revealed substantial variations in reactivities and at least two unusual isotope effects (Fleming *et al.*, 1979; Garner, 1979). Figure 7.4 shows representative data from these studies, in this case for reaction with F_2 as in Eq. [7.3]

$$\mathrm{Mu} + \mathrm{F_2} \rightarrow \mathrm{MuF} + \mathrm{F} \qquad [7.3]$$

at two temperatures. The rate constants and activation energies are given in Table 7.1. Also included in this Table are the rate parameters and isotope effects, where available, for all gas phase muonium reactions published so far. Although the Arrhenius plot (logλ versus K/T) for the F_2 reaction was a good linear plot between 295 and 383 K, represented by $\log k_M = (10.83 \pm 0.20) - (200 \pm 50)/T$, the activation energy is significantly smaller than the H analogue

Figure 7.4. Observed λ for the reaction of Mu with F_2 as a function of F_2 concentration in N_2 gas at 295 K (triangles) and 383 K (squares). The slopes of these lines equal the rate constants k_M, the intercepts correspond to λ_0 (Eq. [7.1]). (Figure taken from Fleming *et al.*, 1979.)

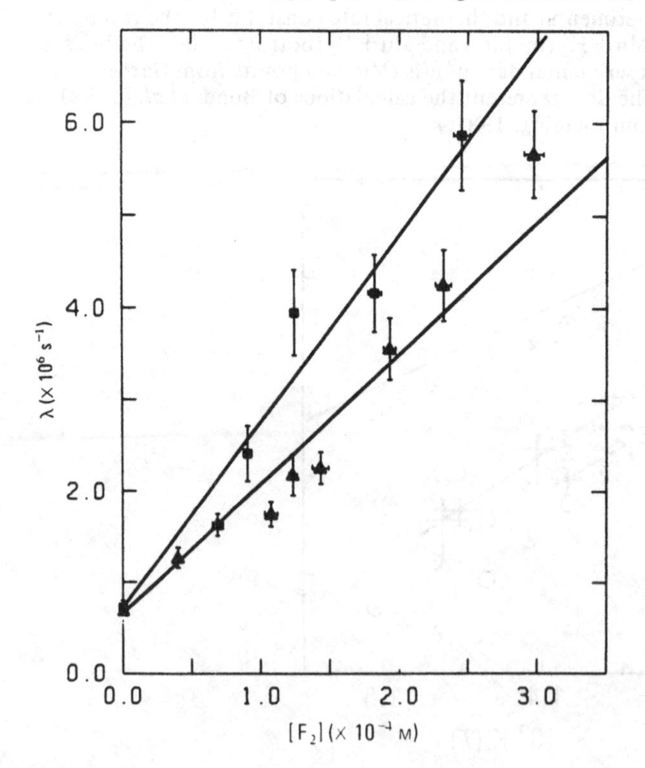

(Fleming *et al.*, 1979). In addition, the isotope effect is considerably greater than the encounter-frequency factor. Both these findings suggest that quantum mechanical tunnelling makes a dominant contribution to this reaction, which agrees precisely with the theoretical predictions of Connor, Jakubetz & Manz (1978). One would normally expect to find upward curvature in the Arrhenius plots (see chapter 8) when tunnelling is important, but perhaps the temperature range was too small for this to manifest itself.

This case of Mu and H reacting with F_2 provides an ideal example in which the lighter isotope is favoured due to tunnelling. It shows, furthermore, the very mass and temperature ranges in which quantum mechanical corrections have to be applied to classical concepts of kinetics. This is a strongly exothermic reaction, so the path taken through the saddle point is an 'early' one (Polanyi & Schrieber, 1974) and the zero-point energy effect is consequently weak. A similar pattern probably prevails in the addition reaction of Mu to C_2H_4, in the high pressure limit. Here there is an isotope effect of 5.8 ± 0.8 (Fleming, Garner & Mikula, 1981) at 295 K, with an activation energy for the H reaction equal to nearly twice that for Mu. In the case of reaction with HCl, on the other hand, the situation is totally different. This reaction is nearly thermo-neutral so the

Table 7.1. *Muonium atom rate constants (k_M) for gas phase reactions at ~295 K; together with activation energies (E_a) and isotope effects (k_M/k_H) where available*

Reactant	$k_M/(10^{10}\,M^{-1}\,s^{-1})$	Reference	$E_a/(kJ\,mol^{-1})$	k_M/k_H [a]
F_2	1.46 ± 0.11	(*a*)	3.9 ± 1.0	9.2 ± 3.1
Cl_2	5.30 ± 0.15	(*a*)	5.7 ± 1.0	3.5 ± 0.8
Br_2	24 ± 3	(*b*)		5.3 ± 1.5
HCl	< 0.0034	(*c*)		< 1
HBr	0.91 ± 0.15	(*a*)	2.9 ± 1.0	3.0 ± 1.0
HI	2.5 ± 0.1	(*c*)		
C_2H_4	0.4 ± 0.05	(*a*)	3.8 ± 1.0	5.8 ± 0.8
	0.55 ± 0.30	(*d*)		
O_2	15.8 ± 2.4	(*e*)	0	2.5 ± 0.4
	11.8 ± 2	(*f*)		
NO	18.3 ± 2.0	(*e*)	0	2.7 ± 0.3
	14.3 ± 3	(*f*)		
C_2H_6	0.04 ± 0.02	(*d*)		
CO_2	0.008 ± 0.004	(*d*)		
CH_3Cl	0.004 ± 0.001	(*d*)		
H_2 at 800 K	0.004 ± 0.0015	(*g*)		
D_2 at 800 K	0.0015 ± 0.0005	(*g*)		

[a] Fleming *et al.*, 1981
[b] Fleming *et al.*, 1976
[c] Fleming *et al.*, 1979
[d] Mobley *et al.*, 1967*a*
[e] Fleming *et al.*, 1980
[f] Mobley *et al.*, 1966
[g] Garner *et al.*, 1982

zero-point energy effect dominates, as in the H_2 reaction. The higher activation barrier causes a 'late' reaction path with the result that the heavier isotope is favoured. In this case there is even an inverse isotope effect ($k_M/k_H < 1$) despite the factor of three in the encounter-frequency term.

Muonium reactions are compared in Table 7.2 with ^1H and ^2H (D) using the limited data available. Perhaps the most striking note to emerge from the overall picture presented in this Table is that k_H/k_D values approximate the encounter-frequency effect of ~1.4 (except for H_2 at high temperature) whereas several of the k_M/k_H values differ substantially from the 2.9 factor for Mu:H. This again underscores one of the principal findings in these Mu studies, namely, that quantum effects take over even at ambient temperatures for such a light atom.

7.3 Theoretical studies of muonium reaction rates

Three of the systems mentioned in the previous section – reaction of Mu with F_2, H_2 and O_2 – have attracted the attention of theorists, because they look like solvable problems, have fundamental issues embodied in them, and the experimental results are there to offer guidance for the next stage of approximation.

The overall temperature-dependence of a reaction rate constant can best be represented in this context (Jakubetz, 1979) by the expression in Eq. [7.4].

$$k(T) = (\pi m^*)^{-1/2}(2/\bar{k}T)^{3/2}\int_0^{\infty} \sigma(E)E\exp(-E/\bar{k}T)\mathrm{d}E \qquad [7.4]$$

In this, m^* is the reduced mass of the reacting particles, which introduces the factor 2.9 for Mu compared to H as the encounter-frequency effect for reactions

Table 7.2. *Comparison of Mu with ^1H and ^2H: ratios of activation energies and rate constants, where available, for gas phase reactions at ~295 K (except for the H_2 reactions which are at 800 K)*[(a)]

Reactant	E_M/E_H[(b)]	k_M/k_H	k_H/k_D
F_2	0.45 ± 0.1	6–14	–
Cl_2	1.2 ± 0.2	3–5	1.5
Br_2	–	4–6	1.5
HBr	0.45 ± 0.15	2.5–4.5	1.8
C_2H_4	0.55 ± 0.15	5–6	1.4
H_2	1.6 ± 0.2	0.055	0.4
O_2	–	2.9	–
NO	–	2.9	–

(a) Data taken from Fleming *et al.*, 1981.
(b) E_M and E_H are, respectively, the energies of activation as determined for the corresponding Mu and H reactions.

of these atoms with relatively heavy molecules. $\bar{k}$ is Boltzmann's constant, E the the total energy, T the temperature, and $\sigma(E)$ is the energy-dependent interaction cross-section. It is this latter term which carries the rest of the isotope effect - the 'dynamic' part - about which the theoretical work revolves.

In order to make a calculation of $\sigma(E)$ there are two major ingredients required, both of which present difficulties even for a very light atom reacting with a homonuclear diatomic molecule. The first is the interaction potential to be used: that is, the energy surface involving the nuclei and their electrons along which the formation of the transition state takes place and which regulates the details of the reaction path. Second, is the nature of the motion itself, the dynamics and rearrangements of the nuclei as they move on that surface. Invariably the Born-Oppenheimer approximation is invoked, as this allows the nuclei and electrons to be treated separately. It also results in a potential surface being formulated which is essentially the same for Mu, ^1H, ^2H and ^3H. It should be borne in mind that this approximation is least valid for the very case in hand (Mu) where the nuclear and electron masses differ by a factor of only 207.

Except for the case of H_3, all potential energy surfaces are based on some degree of empiricism. One of the most common semi-empirical approaches is that originally presented by London-Eyring-Polanyi-Sato and known as the LEPS model (Polanyi & Schrieber, 1974). For the F_2 reaction with Mu, ^1H, ^2H and ^3H, the 'best' extended LEPS potential surface number 2 of Jonathan, Okuda & Timlin (1972) has been adopted (Connor, Lagana, Turfa & Whitehead, 1981). Then, in order to describe the nuclear motion, various types of approaches are possible: based on classical or quasi-classical trajectories, or on quantum mechanical principles. It is easy to imagine that one-dimensional calculations (only valid for collinear encounters) are appreciably simpler than taking all impact angles into account in a three-dimensional calculation. For the F_2 reaction $\sigma(E)$ has been calculated using the following approaches: one-dimensional (collinear) quantum calculations (Connor, Jakubetz & Manz, 1977, 1978; Connor, Jakubetz & Lagana, 1979); quasi-classical trajectories (Cónnor & Lagana, 1979); transition state theory (Connor *et al.*, 1979); vibrating adiabatic models with tunnelling corrections (Jakubetz, 1979*b*); and Franck Condon and impulse energy release models (Jakubetz, 1979*b*; Fischer & Venzl, 1977). Results of the Connor *et al.* (1978) initial study are given as an Arrhenius plot in Figure 7.5, and show the importance of tunnelling in the case of Mu at 300 K in Figure 7.6. These first calculations on the Mu + F_2 reaction preceded the experiments. They predicted the large k_M/k_H and small E_a, almost exactly as the experiments confirmed. Further refinements to the theoretical work continue (Connor *et al.*, 1981). This Mu + F_2 study ranks as a significant case in chemical kinetics in which experiment and theory complemented each other in new terrain.

As a second example, consider the reactions of Mu with H_2 and D_2. Here the potential energy surface does not need to be parametized empirically, since *ab initio* calculations on the intermediate complex (H_3) exist (Lin & Siegbahn, 1978). When the motion on this surface was described using variational transition state theory, the dots in Figure 7.3 were obtained (Bondi *et al.*, 1982). There has also been a classical analysis based on quantum resonances done for this system (Manz, Pollak & Romelt, 1982). Again, there is a strong interdependence of theory with experiment. Again, the agreement is good, especially at the high temperature end of the experimental range for both H_2 and D_2. At the lower temperatures, the reaction was found to be faster than the theory predicts, as if insufficient contributions from tunnelling were included in the calculations.

A third example of the application of theory to muonium reactions in the gas phase concerns the electron spin-exchange interaction with O_2. This is not an activation-controlled reaction and should be quantum mechanical in nature. Such interactions have been successfully calculated when both reactants are monatomic (Berg, 1965; Crampton *et al.*, 1979): indeed, Shizgal (1979) has

Figure 7.5. Arrhenius plots of the collinear calculated rate constants as a function of temperature for the reaction of Mu, ^1H, ^2H (D) and ^3H (T) with F_2 over a broad range of temperatures. The strong upward curvature reminiscent of tunnelling shows in the Mu reaction, particularly at lower temperatures. (Figure taken from Connor *et al.*, 1978.)

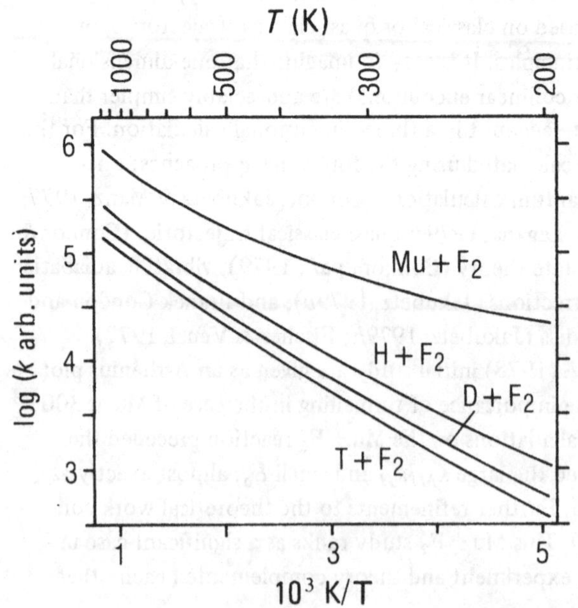

calculated the rate constant for the particular case of Mu interacting with free H atoms. However, for interaction with a rotating, vibrating molecule an exact calculation is more difficult. For the $^{T}Mu + O_2$ spin-exchange, a random phase approximation was first applied within the framework of collision theory (Mobley, 1967*b*); then recently, attempts have been made using both a spherical interaction approximation and an oriented frame decoupling approximation (Aquilanti, Grossi & Lagana, 1981). However, initial results from these latter calculations predict bigger k_M values and considerably larger temperature dependences than are found experimentally for both Mu and H (Mikula *et al.*, 1981).

Figure 7.6. The relative importance of 'tunnelling' and 'classical' interactions are displayed in this figure for the reaction of Mu, H, ^2H (D) and ^3H (T) with F_2 at 300 K. The arrow A indicates the barrier height of 0.0472 eV for the reaction; arrow B is at 0.087 eV, which is the classical threshold for reaction; and C gives kT. (Figure taken from Connor *et al.*, 1978.)

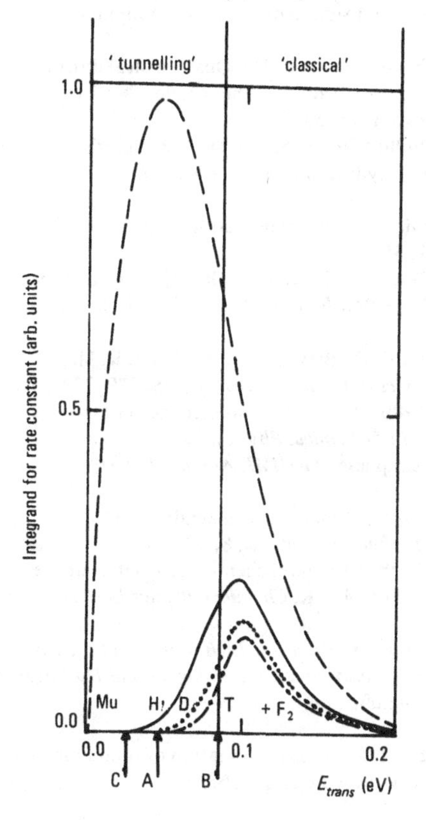

References to chapter 7

Aquilanti, V., Grossi, G. & Lagana, A. (1981). A computational study of spin flip in collisions of H and Mu with oxygen molecules. *Hyperfine Interactions*, **8**, 347–50.

Berg, H. C. (1965). Spin exchange and surface relaxation in the atomic hydrogen maser. *Physical Review*, **A137**, 1621–34.

Bondi, D. K., Clary, D. C., Connor, J. N. L., Garrett, B. C. & Truhlar, D. G. (1982). Kinetic isotope effects in the Mu + H_2 and Mu + D_2 reactions: accurate quantum calculations for the collinear reactions and variational transition state theory predictions for one and three dimensions. *Journal of Chemical Physics*, **76**, 4986–95.

Connor, J. N. L., Jakubetz, W. & Manz, J. (1977). Muonium chemistry: quantum mechanical calculations for the collinear reaction Mu + $F_2(\nu = 0) \rightarrow$ MuF$(\nu' < 3)$ + F. *Chemical Physics Letters*, **45**, 265–70.

Connor, J. N. L., Jakubetz, W. & Manz, J. (1978). Isotope effects in the reaction X + $F_2 \rightarrow$ XF + F (X = Mu, H, D, T): a quantum mechanical and information theoretical investigation. *Chemical Physics*, **28**, 219–30.

Connor, J. N. L., Jakubetz, W. & Lagana, A. (1979). Comparison of quasi-classical, transition state theory, and quantum calculations of rate constants and activation energies for the collinear reaction X + $F_2 \rightarrow$ XF + F (X = Mu, H, D, T). *Journal of Physical Chemistry*, **83**, 73–8.

Connor, J. N. L. & Lagana, A. (1979). Quasiclassical dynamics of light + heavy-heavy atom reactions: the reaction X + $F_2 \rightarrow$ XF + F (X = Mu, H, D, T). *Molecular Physics*, **38**, 657–67.

Connor, J. N. L. (1981). Isotope effects and chemical reaction dynamics of muonium in the gas phase. *Hyperfine Interactions*, **8**, 423–34.

Connor, J. N. L., Lagana, A., Turfa, A. F. & Whitehead, J. C. (1981). Quasiclassical dynamics of light + heavy-heavy and heavy + heavy-light atom reactions: the reaction X + $F_2 \rightarrow$ XF + F (X = Mu, H). *Journal of Chemical Physics*, **75**, 3301–9.

Crampton, S. B., Greytak, T. J., Kleppner, D., Phillips, W. D., Smith, D. A. & Weinrib, A. (1979). Hyperfine resonance of gaseous atomic hydrogen at 4.2 K. *Physical Review Letters*, **42**, 1039–42.

Fischer, S. F. & Venzl, G. (1977). On the dynamics of exothermic triatomic exchange reactions. *Journal of Chemical Physics*, **67**, 1335–43.

Fleming, D. G., Brewer, J. H., Garner, D. M., Pifer, A. E., Bowen, T., Delise, D. A. & Crowe, K. M. (1976) Muonium chemistry in gases: Mu + Br_2. *Journal of Chemical Physics*, **64**, 1281–7.

Fleming, D. G., Garner, D. M., Vaz, L. C., Walker, D. C., Brewer, J. H. & Crowe, K. M. (1979). Muonium chemistry – a review. *Advances in Chemistry Series*, **175**, 279–334.

Fleming, D. G., Mikula, R. J. & Garner, D. M. (1980). Muonium spin-exchange in low pressure gases: Mu + O_2 and Mu + NO. *Journal of Chemical Physics*, **73**, 2751–9.

Fleming, D. G. (1981). Muonium studies in the gas phase. *TRIUMF Annual Report*, pp. 40–2.

Fleming, D. G., Garner, D. M. & Mikula, J. R. (1981). Temperature dependence of muonium reaction rates in the gas phase. *Hyperfine Interactions*, **8**, 337–46.

Garner, D. M., Fleming, D. G. & Brewer, J. H. (1978). Muonium chemistry: kinetics of the gas phase reaction Mu + $F_2 \rightarrow$ MuF + F from 300 to 400 K. *Chemical Physics Letters*, **55**, 163–7.

Garner, D. M. (1979). *Application of the Muonium Spin Rotation Technique to a Study of the Gas Phase Chemical Kinetics of Muonium Reactions with the Halogens and Hydrogen Halides*. Ph.D. Thesis, University of British Columbia, pp. 1–307.

Garner, D. M. *et al.* (1982). Unpublished data.

Hughes, V. W., McColm, D. W., Ziock, K. & Prepost, R. (1960). Formation of muonium and observation of its Larmor precession. *Physical Review Letters*, **5**, 63–5.

Hughes, V. W. (1966). Muonium. *Annual Review of Nuclear Science*, **16**, 445–70.

Jakubetz, W. (1979*a*). Gas phase muonium chemistry, isotope effects, and collision theory: theoretical investigation of the Mu + F_2 and Mu + Cl_2 reactions and their isotopic counterparts. *Hyperfine Interactions*, **6**, 387–95.

Jakubetz, W. (1979*b*). Tuneling in collinear light-heavy-heavy reactions. *Journal of the American Chemical Society*, **101**, 298–307.

Jonathan, N., Okuda, S. & Timlin, D. (1972). Initial vibrational energy distributions determined by infra-red chemiluminescence: III Experimental results and classical trajectory calculations for the H + F_2 system. *Molecular Physics*, **24**, 1143–64. (*Erratum, ibid.*, **25**, 496.)

Lin, B. & Siegbahn, P. (1978). An accurate three-dimensional potential energy surface for H_3. *Journal of Chemical Physics*, **68**, 2457–65.

Manz, J., Pollak, E. & Romelt, J. (1982). A classical analysis of quantum resonances in isotopic collinear H + H_2 reactions. *Chemical Physics Letters*, **86**, 26–32.

Mikula, R. J., Garner, D. M. & Fleming, D. G. (1981). A temperature dependent study of the spin exchange reactions of muonium with O_2 and NO in the range 295 to 478 K. *Journal of Chemical Physics*, **75**, 5362–7.

Mobley, R. M., Bailey, J. M., Cleland, W. E., Hughes, V. W. & Rothberg, J. E. (1966). Muonium chemistry. *Journal of Chemical Physics*, **44**, 4354–5.

Mobley, R. M., Amato, J. J., Hughes, V. W., Rothberg, J. E. & Thompson, P. A. (1967*a*). Muonium chemistry II. *Journal of Chemical Physics*, **47**, 3074–5.

Mobley, R. M. (1967*b*). *Interactions of Muonium with Atoms and Molecules*. Ph.D. Thesis, Yale University, *Dissertation Abstract:* 67-8399, pp. 1–92.

Molin, Yu. N., Salikhov, K. M. & Zamaraev, K. I. (1980). *Spin Exchange*. Berlin: Springer-Verlag.

Pifer, A. E., Bowen, T. & Kendal, K. R. (1976). A high stopping density μ^+ beam. *Nuclear Instrumental Methods*, **135**, 39–46.

Polanyi, J. C. & Schrieber, J. L. (1974). The dynamics of bimolecular reactions. In *Physical Chemistry*, vol. **VIA**, ed. H. Eyring, D. Henderson & W. Jost, pp. 383–487. New York: Academic Press.

Shizgal, B. (1979). A comparison of Mu-H and H-H electron spin exchange cross sections. *Journal of Physics*, **B12**, 3611–17.

8

MUONIUM REACTIONS IN SOLUTION

8.1 Preamble

Although muonium is of intrinsic interest as an 'exotic' atom to those familiar with it, its principal contribution to chemistry will undoubtedly be as a sensitive handle on H-atom processes and in the study of isotope effects. Hydrogen is, after all, our most abundant atom, whereas muonium exists in nature with the utmost rarity – occurring fleetingly at the end of cosmic pion tracks.

Almost all of the liquid phase studies of Mu refer to solutions in water. This is where most of the reference H-atom work exists, and it reflects our aquo-centred existence. Furthermore it is of great practical importance with respect to radiobiology. Free H-atoms are primary reactive species produced in the biosphere as a result of interactions with high-energy ionizing radiations, including natural cosmic and mineral sources as well as man-made nuclear devices. In order to predict, control, and alter the effects of these radiations, it is first necessary to understand the basic chemical reactions involved. But it will be necessary to reach a comprehensive understanding of kinetic isotope effects before muonium can completely usurp hydrogen atom studies for this purpose.

Isotope effects in solution are somewhat different to those outlined for gases (chapter 7, section 1). For instance, the encounter frequency of Mu with S is not given simply by the relative thermal velocities, but is controlled instead by the viscosity of the medium and may be altered by its structure. There will be 'cage' effects, in which many molecular collisions occur for each encounter. Furthermore, the reaction path proceeds via a substantially changed potential energy surface. No longer are the reactants isolated particles: instead, they and their activated complexes are constrained by contact with the solvent molecules through a range of interactions from weak van der Waals or dipolar forces to strong solvation. Polar solvents must have a very marked influence on any reaction involving charged species, at any stage in the process. Finally, particles encounter each other (when at least one is uncharged) either by random diffusion, or by quantum-mechanical tunnelling. The relative importance of each of these will depend on the solvent's viscosity and the particle's mass.

The pioneering work on direct studies of muonium in solution was achieved by Percival and Fischer, and their collaborators, a few years ago (Percival *et al.*, 1976, 1977; Percival, Roduner & Fischer, 1978, 1979). Only ~10% of the incident muons occur as ^{T}Mu in water due to the high diamagnetic yield and the missing fraction (see chapter 6); but, having initially detected this small yield, Percival *et al.* (1976, 1977) immediately set about unveiling its chemistry. They measured k_M with respect to solutes which showed a range of reaction types and isotope effects. These direct observations of Mu by MSR have completely superseded the original P_{res} titration method (as in Figure 4.3) for determining k_M (Firsov & Byakov, 1965; Ivanter & Smilga, 1968; Brewer *et al.*, 1971). Admirable at the time, those indirect model-fitting approaches are fraught with ambiguities, so that very few of them are still used now that direct observations are possible.

Muonium is formed in an inert liquid, such as water, by injecting energetic muons, in an apparatus such as that sketched in Figure 8.1. When working with 24 MeV 'backward' muons, glass cells can be utilized with the solutions deoxygenated by evacuation and sealing. With 'surface' muons at 4.1 MeV their penetration requires very thin-walled cells, as in Figure 8.1 though normally with Mylar windows, so the solutions are deoxygenated by bubbling with an oxygen-free gas (Jean *et al.*, 1978*a*). The water acts as the medium in which Mu is produced and in which it is moderated to thermal energy. Mu has a chemical lifetime in pure water which is at least twice the 2.2 μs natural lifetime of the muon. Substrates with which Mu is to react are usually added at concentrations of 10^{-5} to 10^{-1} M, so they do not intercept Mu formation nor react with it at the epithermal stage. One directly observes the decay of Mu by MSR during the observation time-window of 10^{-7}–10^{-5} s, detecting only the 'triplet' state ^{T}Mu, as described in chapters 3 and 4. This is the ideal pseudo-first-order limit of a bimolecular process in which the exponential decay of the minor component (Mu) is measured in the presence of a constant large excess of the substrate S. Values of k_M come from $(\lambda - \lambda_0)/[S]$ where λ is the decay constant with S at concentration $[S]$ and λ_0 is the 'background' decay constant with $[S] = 0$. For water λ_0 is $\leqslant 2.3 \times 10^5\,s^{-1}$ (Percival *et al.*, 1977; Jean *et al.*, 1978*a*; Nagamine *et al.*, 1982). Typical fitted histograms are shown for phenol in water at 0, 10^{-4} and 5×10^{-4} M in Figure 8.2. A plot of λ versus $[S]$ with typical error bars is given in Figure 8.3, where the slope equals k_M. The fitting procedure often gives standard deviations of 7–15%, but there are other sources of random errors which collectively amount to an uncertainty of about ±25% in most k_M values.

8.2 Mu is neutral

Included in the first few solutes studied (Percival *et al.*, 1977) were those giving exceptionally large (NO_3^- and acetone) and small (aliphatic alcohols) isotope effects. In fact, at one time Mu resembled the hydrated electron as much as it did H in its pattern of reactivity (Jean *et al.*, 1978*a*), as if reaction often proceeded via high Rydberg states or involved electron transfer through the intervening solvent. Therefore, its 'effective charge' at the point of reaction was determined – by studying the primary salt effect on the kinetics of its reactions with positive (Cu^{2+}), neutral (phenol) and negative (SCN^-) solutes. The results of this study are given in Table 8.1. Na_2SO_4 at 0.3 M was used as the inert salt.

These rate data were analysed using the Bronsted-Bjerrum treatment of solutions obeying the Debye-Huckel theory of electrolytes (Hammett, 1970). As the encounter frequency between two species is influenced by coulombic

Figure 8.1. Sketch of the very thin-walled Teflon (now generally Mylar) cell used to study liquids by μSR with weakly penetrating muons. Tc represents a thin-counter (muon start signal), R_1–R_3 and L_1–L_3 are right and left positron counters (stop signal).

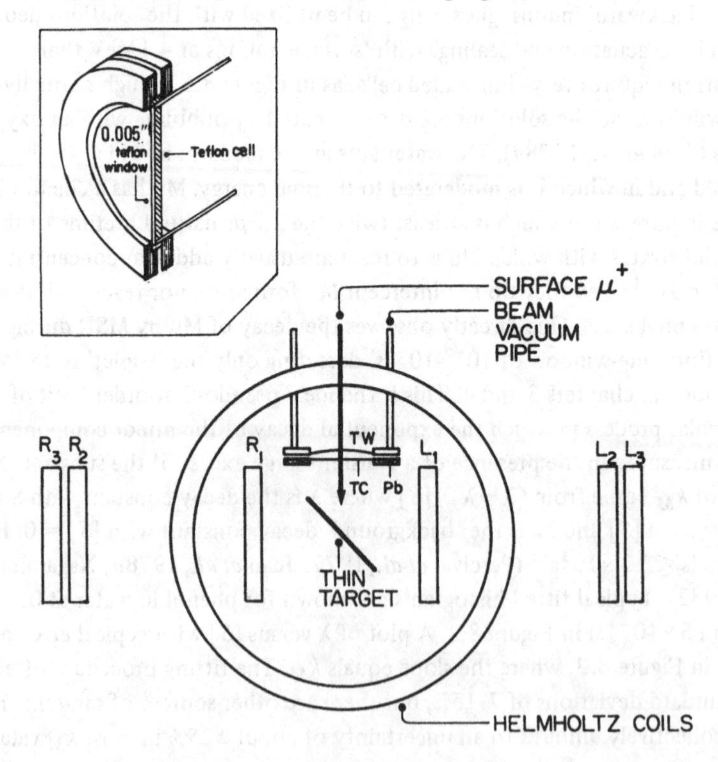

forces arising from their respective charges, the presence of an inert salt creates an ion atmosphere which reduces the encounter frequency between oppositely charged species, increases it between species of the same charge and has no effect when either species is neutral. This is given quantitatively by Eq. [8.1],

$$\log(k/k_0) = \mathrm{A} Z_M Z_s I^{1/2}/(1 + \mathrm{a} I^{1/2}) \quad [8.1]$$

Figure 8.2. Representative muonium precession signals at 7.6 G in (a) pure water, (b) 10^{-4} M phenol, and (c) 5×10^{-4} M phenol in aqueous solution at ~295 K. The points give the binned histogram with error bars shown every 10. Total of ~10^7 events in each histogram. The line is the computer fit after removal of the decay due to the 2.2-μs muon lifetime and the background. Decay of the high-frequency muonium precession, superposed on a slow muon precession, can readily be seen to occur faster the higher the phenol concentration. (Figure taken from Jean *et al.*, 1978*a*.)

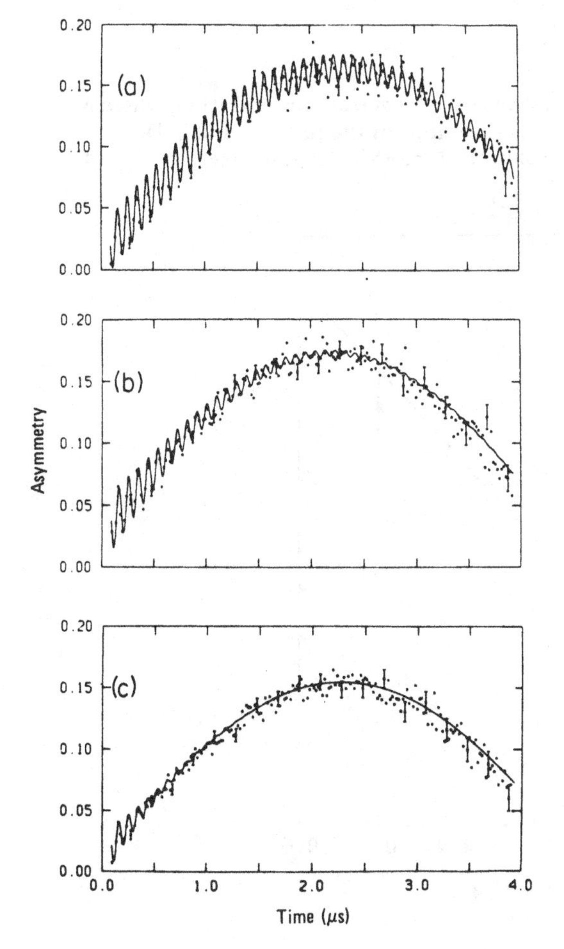

Table 8.1. *Muonium rate constants obtained for solutes in aqueous solution at small and large ionic strengths, and the effective charge on Mu calculated therefrom (Z_M) (Jean* et al., *1979)*

	k_M ($10^8\,M^{-1}s^{-1}$)			
Solute	($I<0.02$)	($I=0.9$)	k/k_0[(a)]	Z_M[(b)]
Cu^{2+}	65 (±1)	63.5 (±1.1)	0.98 (±0.23)	−0.01 (±0.11)
C_6H_5OH	72 (±4)	65 (±4)	0.90 (±0.12)	(c)
SCN^-	0.62 (±0.06)	0.74 (±0.04)	1.19 (±0.13)	−0.15 (±0.10)

(a) Ratio of k_M values at $I<0.02/I=0.9$
(b) Value of Z_M calculated using Eq. [8.1]
(c) Assuming phenol is not ionized significantly at 2×10^{-4} M then $Z_s=0$ so Z_M is indeterminable.

Figure 8.3. Representative plot of λ versus concentration, where λ_0 is the intercept shown and the slope of the line equals k_M. Data refer to phenol in aqueous solution at ~295 K. (Figure taken from Jean *et al.*, 1978*a*.)

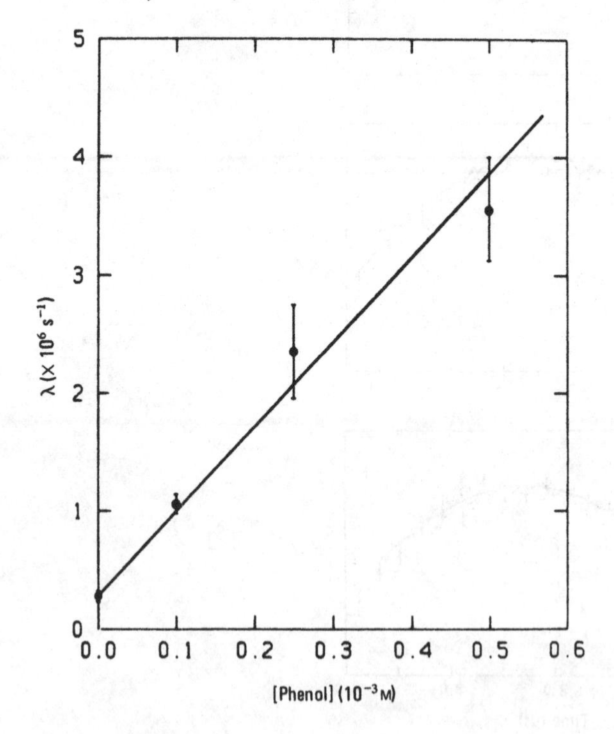

where k is the rate constant in the presence of total ionic strength $I(=0.5\Sigma c_i z_i^2)$, k_0 the rate constant at $I \to 0$, Z_M is the charge on muonium, Z_s the charge on the solute, with A and a as two solvent-dependent constants numerically equal to 1.02 and 1.0, respectively, for water. The values calculated for Z_M are given in the last column of Table 8.1, where it can be seen that the charge on Mu is not significantly different from zero. This means that even for an efficient (nearly diffusion-limited) spin-exchange reaction, such as with Cu^{2+}, there is probably no long-range charge separation or electron transfer: the process occurs only after Mu and Cu^{2+} have formed an ordinary encounter complex.

8.3 Basic pattern of reactivity of Mu

The main characteristics of muonium chemistry are highlighted by the selection of reactions given in Table 8.2, and by the comparison of rate

Table 8.2. *Some selected muonium rate constants (k_M)[(a)] for a variety of reaction types, and their comparison with H-atoms (k_H)[(b)], hydrated electrons (k_e)[(c)] and positronium (k_p)[(d)], where available. (All data refer to aqueous solutions at ~295 K[(e)])*

Reaction type	Solute	k_M ($M^{-1}s^{-1}$)	k_M/k_H	k_M/k_e	k_M/k_p
Reduction	MnO_4^-	2.5×10^{10}	1.0	1	1
	Ag^+	1.6×10^{10}	0.5	0.5	>10^3
	Tl^+	8×10^8	7	0.03	>8
	Cd^{2+}	8×10^6	–	0.0002	–
Addition	maleic acid	1.1×10^{10}	1.4	0.3	–
	acrylamide	1.9×10^{10}	1	1	–
	p-nitrophenol	8×10^9	2.7	0.3	1
	phenol	7×10^9	3.5	400	>70
Abstraction	methanol	3×10^4	0.01	>3	–
	2-propanol	5×10^5	0.01	–	–
	HCO_2^-	8×10^6	0.03	800	–
Others	OH^-	1.7×10^7	1.0	>10^3	–
	NO_3^-	1.5×10^9	155	0.15	–
	$(CH_3)_2CO$	8.7×10^7	40	0.01	–
	Ni^{2+}	1.7×10^{10}	–	1	–
	I_2	1.7×10^{10}	1	1	1
	O_2	2.4×10^{10}	2	1	1

(a) k_M values taken from Walker, 1981.
(b) k_H values taken from Anbar *et al.*, 1975.
(c) k_e values taken from Anbar *et al.*, 1973 and supplement.
(d) k_p values taken from Ache, 1979, or Goldanskii, 1968.
(e) No error limits are given and the values for the ratios often have considerable uncertainty (~50%).

constants of Mu with H, e^-_{aq} and Ps (where possible). Different types of reactions can be expected to show different isotope effects, because of the change in mechanism. In fact, k_M/k_H varies from ~10^{-2} to 10^2. This 10^4-fold change even shows up merely by changing the solute from an alcohol to a ketone. In solution there is also the natural upper limit to reaction rates set by the diffusion-limited encounter frequency.

Among the reducible solutes of the first group are two powerful oxidizing agents, MnO_4^- and Ag^+, which react with Mu and its cohorts essentially at the diffusion-controlled limit, and no significant isotope effect apparently results (see later). With Tl^+ and Cd^{2+}, however, there is a full range of reactivity. It is interesting that both these ions are readily reduced by e^-_{aq} though reactivity with Mu has fallen off; so this implies that their reduction potentials lie in the neighbourhood of Mu and significantly less negative than the −2.7 V of e^-_{aq} (Baxendale, 1964). Cd^{2+} is a rare solute in that it reacts some 10^4 times faster with e^-_{aq} than with Mu (see chapter 6 section 4).

Of the addition reactions, those solutes with isolated vinyl bonds (maleic acid and acrylamide) react at about the encounter rate with Mu, H and e^-_{aq}. The contrast between *p*-nitrophenol and phenol is interesting, however, because the former reacts efficiently with all four species, while phenol reacts rapidly only with Mu and H. The electromeric effect of the nitro group has no influence in this case on Mu or H, whereas it was probably responsible for reaction by e^-_{aq} and Ps with *p*-nitrophenol. Ps reactions are notoriously sensitive to nitro-groups (Ache, 1979).

Abstraction reaction efficiencies have a strong mass dependence, and may be influenced by the actual free energy of the reaction (see later). H abstracts hydrogen atoms much more efficiently than Mu, which in turn is much better than e^-_{aq}. With bromine-substituted compounds the reaction with Mu was fast and k_M/k_H was found to be ~1 (Percival *et al.*, 1979); but this may be dissociative electron transfer rather than abstraction.

Further indications of muonium's overall reactivity picture are contained in Table 8.2. (i) It is interesting that μ^+-transfer to OH^- occurs with the same efficiency as proton transfer from H, despite the proton- and hydroxyl-dominated medium. Furthermore, the Mu + OH^- reaction has the temperature-dependence of a normal bimolecular reaction in water (Ng *et al.*, 1982). Hydrated electrons do not react with OH^- because e^-_{aq} is the fully basic state – being already the conjugate base of H. (ii) Reactions of Mu with NO_3^- and acetone seem odd at this time – except that they place Mu between H and e^-_{aq}. (iii) With Ni^{2+} (a d^8 ion) there is an efficient spin-conversion (possibly spin-exchange) interaction with Mu transforming $^TMu \rightarrow {}^SMu$ (see later), the analogues of which are not observable in the methods used to measure H and

e_{aq}^- reactions. However, Ni^{2+} is a system like Cd^{2+} whose one-electron reduction occurs very readily by e_{aq}^- but not by H (and, presumbly, not by Mu). (iv) I_2 reacts with all four species at about the same rate, probably the diffusion-limited rate. For H this is presumed to be an abstraction or substitution reaction (indistinguishable in this case) whereas with e_{aq}^- it is an electron-transfer followed by dissociation. (v) O_2 also reacts at the maximum rate with all four species. This means that the O_2 content of all solutions in which Mu is to be studied by MSR must be reduced to $\leqslant 10^{-5}$ M. This level of deoxygenation of water is readily achieved by reducing the air-saturated concentration by a factor of $\geqslant 20$ either by bubbling with an oxygen-free gas for several minutes or by a few freeze-pump-thaw cycles. With regard to the Mu + O_2 reaction mechanism, there are several possibilities - simple combination, electron-transfer reduction, or spin-exchange. They can all occur together. But, when an overall (observed) rate constant is at the diffusion-controlled limit, it matters not how many competitive alternative reaction paths are present: the maximum rate of loss of Mu is already set by the rate at which encounters occur.

8.4 Diffusion-limited rates

Muonium reactions in solution (Ng *et al.*, 1981) have been represented by the commonly used scheme in Eq. [8.2],

$$\mathrm{Mu} + \mathrm{S} \underset{k_b}{\overset{k_d}{\rightleftarrows}} [\mathrm{Mu} \ldots \mathrm{S}] \xrightarrow{k_p} \mathrm{P} \qquad [8.2]$$

where [Mu . . . S] is the encounter-pair and P the product of reaction. Being caged by the solvent, the encounter pair has a substantial lifetime during which many collisional activations can occur (perhaps $\sim 10^2$ in $\sim 10^{-11}$ s) before either producing P or dissociating. Since Mu is neutral the initial encounter occurs by random diffusion with a rate characterized by the bimolecular rate constant k_d; whereas first order rate constants k_p and k_b characterize the decay rates of the encounter-pair. The overall rate of reaction is given by Eq. [8.3],

$$-\mathrm{d}[\mathrm{Mu}]/\mathrm{d}t = k_{obs}[\mathrm{Mu}]\,[\mathrm{S}] = k_p[\mathrm{Mu} \ldots \mathrm{S}] \qquad [8.3]$$

which gives Eq. [8.4]

$$k_{obs} = k_p k_d/(k_p + k_b) \qquad [8.4]$$

when one applies the steady-state approximation to [Mu . . . S]. (In μSR the different states are observed one at a time with a distribution coming from summation over time; but this is equivalent to summing over all states at one time.)

There are two simplifying limits to Eq. [8.4]. (i) When encounters seldom lead to reaction ($k_p \ll k_b$) then the observed rate constant, k_{obs}, equals $k_p k_d/k_b$. This corresponds to an activation-controlled reaction. The Arrhenius parameters are connected to the free-energy change of a pseudo-preequilibrium intermediate complex. (ii) When almost all encounters lead to reaction ($k_p \gg k_b$) then $k_{obs} = k_d$. This is the diffusion-controlled limit. Here the overall reaction rate is determined merely by the encounter rate. In general k_d can be represented by interaction radii (R) and diffusion coefficients (D), as in Eq. [8.5] for Mu + S,

$$k_d = (R_s + R_M)(D_s + D_M)(4\pi \mathrm{N_A} 10^{-3}) \qquad [8.5]$$

where $\mathrm{N_A}$ is Avagadro's number and R and D are in cgs units. Muonium is usually so much smaller and more mobile than S that Eq. [8.5] approximates to [8.6],

$$k_d \sim R_s D_M (4\pi \mathrm{N_A} 10^{-3}) \qquad [8.6]$$

so that the diffusion-limited rate is seen to depend almost entirely on the size of S and the diffusion of Mu. Any temperature-dependence will lie with D_M.

Muonium reaction rates in aqueous solution have been studied as a function of temperature over the range of liquid water for five different reaction types by Ng *et al.* (1981). Arrhenius plots are shown in Figure 8.4 and the parameters A and E (defined by: $\log k = \log A - E/2.3RT$) are provided in Table 8.3 for these five reactions. A common value of E, equal to $17 \pm 2\,\mathrm{kJ\,mol^{-1}}$, has been

Figure 8.4. Arrhenius plots for the reactions of Mu with: hollow triangle, Ni^{2+}; filled circle, MnO_4^-; hollow circle, maleic acid; filled triangle, NO_3^-; and filled square, HCO_2^-. (Figure taken from Ng *et al.*, 1981.)

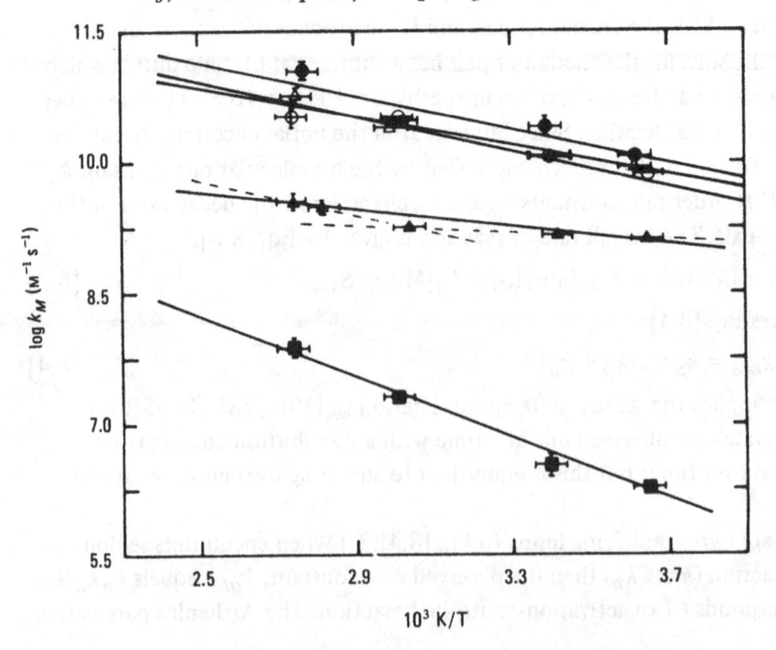

found for the three reactions having $k_M > 10^{10}\,\mathrm{M}^{-1}\,\mathrm{s}^{-1}$, namely, reduction of MnO_4^-, addition to maleic acid (*cis*($CHCO_2H)_2$), and spin-conversion with Ni^{2+}. This value of E is approximately equal to the value often found for diffusion-limited reactions in water (Pilling, 1975), and to ΔE_{visc}, the energy barrier to elementary flow processes obtained for the viscosity (η) using Eq. [8.7],

$$\eta = B\exp(\Delta E_{visc}/RT) \qquad [8.7]$$

where B is a parameter independent of temperature. Indeed, following Lazzarini & Lazzarini (1980), it can be shown that k_M/T is approximately inversely proportional to η for these three fast reactions. All of this points to the fact that these three reaction rates are at the diffusion-controlled limit.

It transpires that the fastest 15 of the known 47 rate constants of muonium in aqueous solution all lie in the narrow range $(1.1 \text{ to } 2.6)\times 10^{10}\,\mathrm{M}^{-1}\,\mathrm{s}^{-1}$ (see Appendix). This suggests that they all occur on nearly every collision and that their variation arises mainly from different interaction radii (R_s). But before trying to correlate the A factors with R_s one would need to know the role of ligands and solvation shells in Mu reactions. At least there is no possibility of a time-dependence associated with these k_d values, since they are measured over $\sim 10^{-6}$ s (Walker, 1981). It should also be noted that in arriving at the diffusion-limited approximation of Eq. [8.4], a wide range of k_p values are covered by an encounter, due to the caging effect of the solvent. One obtains the same k_d value for all reactions in that limit, regardless of whether reaction occurred on first collision or whether a relatively long-lived [Mu . . . S] complex was necessitated by a reaction of relatively low intrinsic probability.

Table 8.3. *Arrhenius parameters for five types of muonium reactions in aqueous solution*[a]

Solute (reaction type)	k_M ($\mathrm{M}^{-1}\mathrm{s}^{-1}$)	A ($\mathrm{M}^{-1}\mathrm{s}^{-1}$)	E (kJ mol^{-1})	k_M/k_H[b]
MnO_4^- (reduction)	2.5×10^{10}	3.5×10^{13}	18 (±2)	1.0 (±0.3)
Maleic acid (addition)	1.1×10^{10}	2.3×10^{13}	19 (±2)	1.4 (±0.4)
Ni^{2+} (spin-conversion)	1.7×10^{10}	0.9×10^{13}	16 (±2)	(*c*)
HCO_2^- (abstraction)	8×10^{6}	0.4×10^{13}	33 (±2)	0.03
NO_3^- (?)	1.5×10^{9}	2×10^{10}[d]	6 (±1)[d]	155 (±20)

[a] Data from Ng *et al.*, 1981. k_M refers to ~295 K.
[b] k_H data selected from the Tables of Anbar *et al.*, 1975.
[c] No spin-conversion data available for H.
[d] These values were obtained from the least-squares best straight line. In fact the data fall on a smooth shallow curve in the Arrhenius plot of Figure 8.4, with the slope at the high T end giving $A = 3\times10^{12}$ and $E = 20$ (±2), at a low T end $A = 6\times10^{8}$ and $E = 3.5$ (±1.5).

For all Mu reactions with $k_M > 10^{10}\,\mathrm{M}^{-1}\,\mathrm{s}^{-1}$ the mean value of k_M/k_H is found to be 1.4 ± 0.5 (this excludes spin-conversion for which H analogues are unavailable, and I_3^- whose value is now suspect (Stadlbauer, Ng, Jean & Walker, 1982)). Thus, there is essentially no kinetic isotope effect in the diffusion-controlled limit of these reactions' rates in aqueous solution. There is not even the factor of three for the higher mean velocity of Mu.

If Eq. [8.5] is valid this means that the diffusion coefficients of H and Mu are essentially equal. In turn this means that diffusion does not depend on molecular velocity, even for species, such as these, which are much smaller than the cavities in the solvent structure. The fact that Mu 'bounces about' with three times the ferocity of H, apparently does not enhance its chance of escaping to the next cavity. This also means that diffusion in water is not governed by quantum-mechanical tunnelling at these temperatures, even for an atom as light as Mu, since tunnelling has a strong mass dependence.

When Eq. [8.5] is combined with the Stokes-Einstein formulation of diffusion coefficients (Eq. [8.8]),

$$D = \bar{k}T/R\pi\beta\eta \qquad [8.8]$$

(where $\bar{k}$ is the Boltzmann constant and β is a parameter which varies only slightly with the relative sizes of solute and solvent molecules) one finds that k_d depends only on radii, as in Eq. [8.9].

$$k_d \propto (R_s + R_M)^2/R_s R_M \qquad [8.9]$$

k_M/k_H ratios of unity are consistent with Eq. [8.9] for a given solute because Mu and H have the same Bohr radii. So, isotope effects disappear when one invokes the Stokes-Einstein relationship (though R in Eq. [8.5] is an interaction radius, whereas R in Eq. [8.8] is a physical radius governing flow). This conclusion seems perfectly reasonable for large atoms which have to push the solvent aside when they move, because the Stokes-Einstein relationship represents bulk hydrodynamic properties. One would have thought such bulk properties were less important for atoms as small as H, however.

Diffusive motion seems to be dominated by solvent structure, where the physical size of the solute counts, rather than its mass and velocity. This implies that diffusion is controlled by the collective motion of the solvent through dislocation fluctuations and cavity migration. As it happens, the diffusion-limited rates of e^-_{aq}, and Ps are also in the same range, (1 to 3) $\times 10^{10}\,\mathrm{M}^{-1}\,\mathrm{s}^{-1}$, in water at ~295 K (Hart & Anbar, 1970; Nichols, Wild, Bartal & Ache, 1974; Anbar, Bambenek & Ross, 1973; Jean, Fleming, Ng & Walker, 1979*b*). So the same factors which dominate diffusion of Mu and H apply equally to solvated charges like e^-_{aq} and to super-light quantum species like Ps. Perhaps this is peculiar to water with its hydrogen-bonded and ordered structure. In hydrocarbons this is unlikely to be so because the mobility of solvated electrons is at least 10^2-fold higher than in water (Allen, 1976) and varies by a further three orders of magnitude depending on such properties as the sphericity and polarizability of the solvent molecules (Dodelet & Freeman, 1972). It will be interesting therefore to see how k_M and k_M/k_H vary with viscosity in hydrocarbon solvents.

8.5 Activation-controlled reactions

Arrhenius parameters for the reaction of Mu in an abstraction reaction with HCO_2^- (Table 8.3) strongly imply that this is a simple activation-controlled reaction. The A-factor is comparable to the rate of encounter, but there is a large activation energy of 33 kJ mol^{-1}, and an even larger one for the Mu + DCO_2^- reaction (Ng *et al.*, 1982). These reactions have inverse isotope effects strongly favouring abstraction by the heavier atom. Table 8.4 in fact shows a continuous trend in rate constants as the ratio of the masses of the abstracting/abstracted atom changes. This trend can be qualitatively understood in terms of simple transition-state theory by the zero-point energy changing the height of the vibrationally-adiabatic barrier and by the overall free energy change (Ng *et al.*, 1981, 1982).

Inverse isotope effects were also found for the abstraction of hydrogen by Mu in neat methanol and other alcohols at various temperatures, including low ones (Percival *et al.*, 1979; Roduner, 1979; Roduner & Fischer, 1979). These authors applied transition state theory, incorporating possible tunnelling for a Mu + S reaction using Eq. [8.10],

$$k = \Gamma(\bar{k}T/h)Q^{\dagger}/Q_M Q_S)\exp(-E/RT) \quad [8.10]$$

where Γ is the transmission factor which includes the effect of tunnelling, h is Planck's constant, $Q^{\dagger}$, Q_M and Q_S are the molecular partition coefficients

Table 8.4. *Effect of mass ratio of abstracting/abstracted atom on the rate constant for abstraction from formate ions in aqueous solution at ~295 K*

Reaction	k (M^{-1} s^{-1})	$M_1/M_2$$^{(a)}$	Reference
Mu + DCO_2^-	1.1×10^6	0.056	(*b*)
Mu + HCO_2^-	3.4×10^6	0.11	(*c*)
H + DCO_2^-	2.3×10^7	0.50	(*d*)
H + HCO_2^-	1.2×10^8	1.0	(*d*)

(*a*) M_1 is the mass of the abstracting atom (Mu or H), M_2 that of the abstracted atom (H or D).
(*b*) Ng *et al.*, 1982.
(*c*) Ng *et al.*, 1981.
(*d*) Anbar *et al.*, 1975.

of the activation complex, Mu and S, respectively. On comparing Mu with H one gets Eq. [8.11],

$$k_M/k_H = (\Gamma_M/\Gamma_H)(Q_H/Q_M)(Q^\dagger_{MX}/Q^\dagger_{HX})\exp\{(E^\dagger_{HX} - E^\dagger_{MX})_{vib}/RT\} \times \exp\{(E_H - E_M)_{el}/RT\} \quad [8.11]$$

where the difference in activation energies $(E_M - E_H)$ has been factored into vibrational (vib) and electronic (el) components. Dealing qualitatively with each of the five terms on the right-hand side of Eq. [8.11], one can see the following: the first term is >1, being considerably greater than unity if tunnelling plays a dominant role; the second term lies between 1 and 27 (it would be 27, proportional to $m^{3/2}$, in the gas phase; but is much less and probably close to unity in the liquid due to the miniscule free volume); the third term is approximately unity since S is much more massive than H or Mu, and the vibrational component in $Q^\dagger$ is minor at room temperature; the fourth term represents the difference in zeropoint vibrational energy of the H and Mu activated complexes, so this is a factor less than unity; finally, the fifth term is approximately unity since the electronic energies are equal within the Born-Oppenheimer approximation (in reality a touch greater than one).

In view of the fact that hydrogen abstraction reactions are found to have a large inverse isotope effect (see also Table 8.2) it follows that the fourth term – the zero-point energy of the activated complexes – must be of overriding dominance. Indeed, when bond-energy bond-order (BEBO) calculations were performed the magnitude of the fourth term was found to be basically in accord with the trend in isotope effects. This suggested tunnelling plays a minor role (Percival *et al.*, 1979). However, tunnelling factors were also calculated using the energy barriers shown in Figure 8.5 with the vibrational contributions to both height and width included (Roduner, 1979; Roduner & Fischer, 1979). The authors were then able to conclude that hydrogen abstraction by Mu occurs mainly by tunnelling; and that one can still get an inverse isotope effect for reactions which are not too exothermic, because of the shape of the barriers.

8.6 Reactions possibly modified by tunnelling

For the reaction of Mu with NO_3^- one finds both Arrhenius parameters to be exceptionally small (see Table 8.3). E is much smaller even than the activation energy for diffusion; while A is three orders of magnitude less than the encounter frequency, despite there being no obvious orientational or steric problems. When these facts are coupled with the observed small upward curvature in the Arrhenius plot (see Figure 8.4), one inclines to the view that tunnelling may be involved here. Although the curvature is barely beyond the error limits and the temperature range is rather small, this Arrhenius plot of the NO_3^- reaction has the appropriate characteristics expected for tunnelling and with a reasonable high temperature slope (see footnote (*d*) to Table 8.3). One type of curvature is demonstrated in the sketch of Figure 8.6.

Tunnelling is of considerable interest in its own right, but it can distort kinetic isotope effects beyond all recognition. Indeed, the NO_3^- reaction has a k_M/k_H value of 155 (±20). Tunnelling depends on mass, barrier shape (width and height) and free energy release. It will be manifest at room temperature for Mu only with the optimum reaction barrier: perhaps NO_3^- is a case in point. However, curvature like that in Figure 8.6 in Arrhenius plots can also arise from the presence of alternative (competitive) reaction paths having different *A* and *E* values but comparable rates at the mid temperature: in fact, just as in footnote (*d*) values of Table 8.3.

8.7 Spin-conversion (spin-exchange) reactions

Jean, Brewer, Fleming & Walker (1978*b*) studied a wide range of transition metal ions in aqueous solution, and found there to be strong preference for reaction (depolarization of ^{T}Mu) with ions which were paramagnetic. The relevant rate data are given in Table 8.5. For instance, consider the interaction of Mu with the d^6 ion Fe(II). When present as Fe^{2+}_{aq} in the weak ligand

Figure 8.5. Potential barriers for the reactions of Mu and H with methanol. Dashed line, modified BEBO potential along a skewed, mass-weighted coordinate system; full line, the vibrational adiabatic energy. (Figure taken from Roduner & Fischer, 1979.)

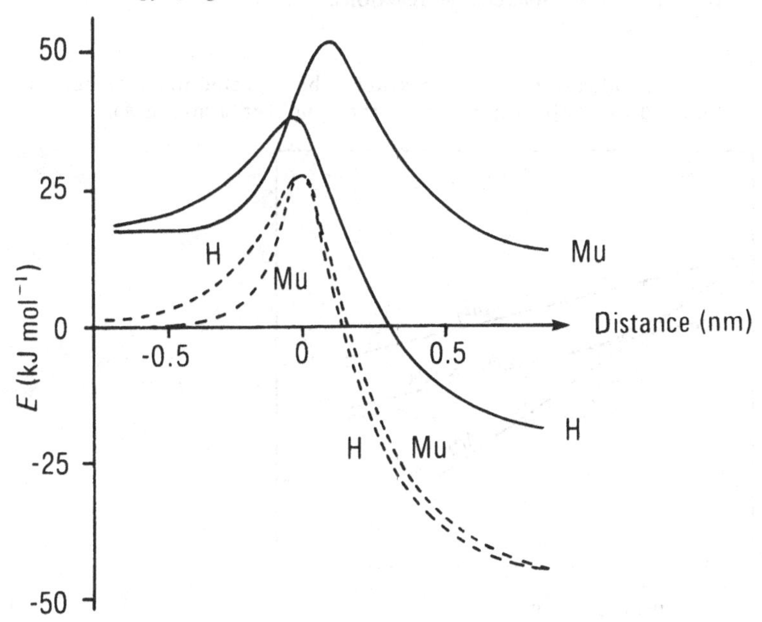

fields of SO_4^{2-} or H_2O it is paramagnetic and interacts with ^{T}Mu at about the diffusion-limited rate. But when present in the strong ligand fields of $Fe(CN)_6^{4-}$, where it is low spin and diamagnetic, its reaction rate with Mu is 40-times slower. Or compare $Fe(CN)_6^{3-}$ with $Fe(CN)_6^{4-}$, where the paramagnetic ion reacts 70-times faster. (Here chemical reduction of Fe(III) could be contributing to

Table 8.5. *k_M values for paramagnetic and diamagnetic solutes in aqueous solution at ~295 K*[(a)]

Solute	k_M (10^{10} M^{-1} s^{-1})	Spin[(b)]	k_H (M^{-1} s^{-1})[(c)]
Fe^{2+}	1.2	2	2×10^7
$Fe(CN)_6^{4-}$	0.03	0	–
$Fe(CN)_6^{3-}$	2.0	1/2	2×10^9
Zn^{2+}	$<10^{-3}$	0	$<3 \times 10^5$
Ni^{2+}	1.7	1	$<3 \times 10^5$
$Ni(cy)(NH_3)_2^{2+}$	2	1	$<3 \times 10^5$
$Ni(cy)^{2+}$	0.02	0	$<3 \times 10^5$
Cr^{3+}	0.53	3/2	2×10^7
Cu^{2+}	0.65	1/2	7×10^7
Cd^{2+}	10^{-3}	0	$<3 \times 10^5$

(a) Data taken from Jean *et al.*, 1978*b*, except the Nickel cyclam (cy) data – taken from Stadlbauer *et al.*, 1982.
(b) Total electron spin quantum number.
(c) Mean or best values in Tables from Anbar *et al.*, 1975. None of these refer to spin-exchange reactions.

Figure 8.6. Indication of the curvature to be expected in an Arrhenius plot caused by contributions from tunnelling (for common *A*).

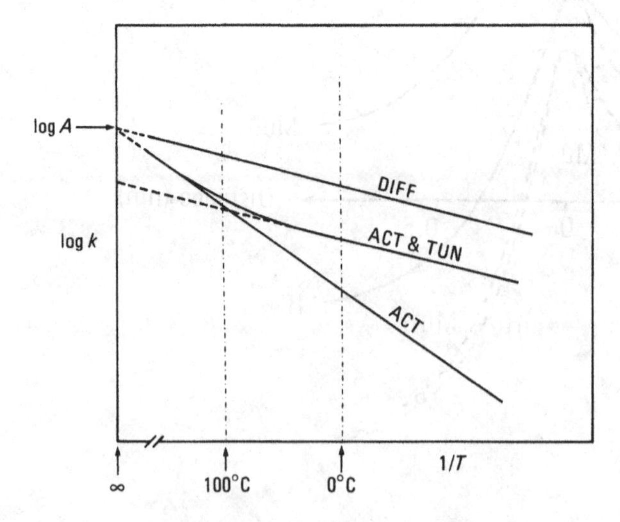

that rate, but this may be a relatively small effect since k_H is only $2 \times 10^9 \text{M}^{-1}\text{s}^{-1}$ (Anbar, Farhataziz & Ross, 1975).) One can equally compare Ni^{2+} (paramagnetic, with $k_M = 1.7 \times 10^{10} \text{M}^{-1}\text{s}^{-1}$) with Zn^{2+} (diamagnetic and $k_M < 10^7 \text{M}^{-1}\text{s}^{-1}$), or paramagnetic Cr^{3+} and Cu^{2+} ions ($k_M \sim 5 \times 10^9 \text{M}^{-1}\text{s}^{-1}$) with diamagnetic Cd^{2+} ($k_M < 10^7 \text{M}^{-1}\text{s}^{-1}$).

In each case there is a greatly increased rate of loss of ^{T}Mu with paramagnetic ions. This is almost certainly an electron spin-conversion process, transforming $^T\text{Mu} \rightarrow {}^S\text{Mu}$ with loss of muonium precession and depolarization of the muon spin. It is not known at this stage whether such reactions actually involve electron spin-exchange (though that is often mentioned: Molin, Salikhov & Zamaraev, 1980) or merely catalysed spin-flip.

Direct evidence for a spin-conversion reaction with O_2 and NO was provided by the use of longitudinal fields for the gas phase studies (Mobley *et al.*, 1967, see Figure 7.2). Although this has not been used in the liquid phase reactions there are two strong pieces of evidence supporting the spin-conversion assignment. First, is the fact that Ni^{2+} at ~ 1 M concentration does not enhance P_D (see chapters 4 and 6, and Table 6.5). This means the reaction between Mu and Ni^{2+} is not a chemical reduction (electron transfer) but it could be spin-conversion. Second, when the spin-state of Ni^{2+} was changed from $S = 0$ to $S = 1$ without altering its oxidation state or the nature of its principal ligands, then the value of k_M increased some hundred-fold. This was achieved by using nickel cyclam (1,4,8,11-tetraazacyclotetradecane nickelate II) which has the rare property of switching from an elongated octahedron (paramagnetic, $S = 1$) to a square-planar configuration (diamagnetic, $d_{x^2-y^2}$ unoccupied) merely by the addition of an inert salt such as $NaClO_4$ or Na_2SO_4. Whereas one cannot in practice do the measurement in 100% of either of these states, their proportion was changed by inert salt and the compositions evaluated by means of their optical absorption spectra. Figure 8.7 shows the variation of k_M with the fraction of paramagnetic component. One can see that with only diamagnetic species present $k_M < 5 \times 10^8 \text{M}^{-1}\text{s}^{-1}$. At the other end, with high proportions of paramagnetic species, it looks as if the transition from two axial H_2O to NH_3 ligands caused the plateauing effect, and that k_M was $\sim 4 \times 10^{10} \text{M}^{-1}\text{s}^{-1}$ with H_2O ligands and $\sim 2 \times 10^{10} \text{M}^{-1}\text{s}^{-1}$ with NH_3 ligands. This vast enhancement in k_M with conversion from diamagnetic to paramagnetic complex provides direct corroboration of its assignment as a spin-conversion process.

Other matters of interest with regard to these spin-conversion reactions of Mu are the following. First, the rate constants are essentially at the diffusion-controlled limit: therefore they result from an ordinary encounter and are not long-range effects. This is clear from the Arrhenius parameters for the Ni^{2+} reaction in Table 8.3, as discussed in section 3 above, and spin-exchange reactions are closest to outer-sphere electron-transfer reactions. Second, no correlation was found between the actual value of k_M and the number of unpaired electrons. But no simple correlation would be expected in the diffusion-controlled limit. Third, this fast spin-conversion of $^TMu \rightarrow {}^SMu$ may find use as a probe of the spin state, and hence of ligand field strengths, for general use in chemistry. It would be a non-destructive way to decide about d^6 ions of Fe^{2+} and Co^{3+} in various complex biological systems if other techniques prove to be unsuitable. Fourth, one is not bothered at all when using the μSR method by the possible occurrence of the reverse process, conversion from ^SMu to ^TMu. This is because once the 'singlet' state is formed it immediately depolarizes the muon spin by hyperfine oscillations. Therefore, wherever a $^SMu \rightarrow {}^TMu$ conversion does occur, the ^TMu formed is not observable because it has been spin depolarized.

8.8 Muonium reactions with organic molecules

Most organic molecules are sufficiently soluble in water to give millimolar concentrations in solution, so the values of k_M for any relatively fast reaction with Mu can be studied. Many have been, as the Appendix shows.

Figure 8.7. Plot of k_M against the fraction of nickel cyclam in the paramagnetic form. The two points on the right were obtained with NH_3 as the axial ligand, the other three have H_2O as axial ligand. Changes in composition were caused by the addition of Na_2SO_4, $NaClO_4$ or NH_3. (Figure taken from Stadlbauer *et al.*, 1982.)

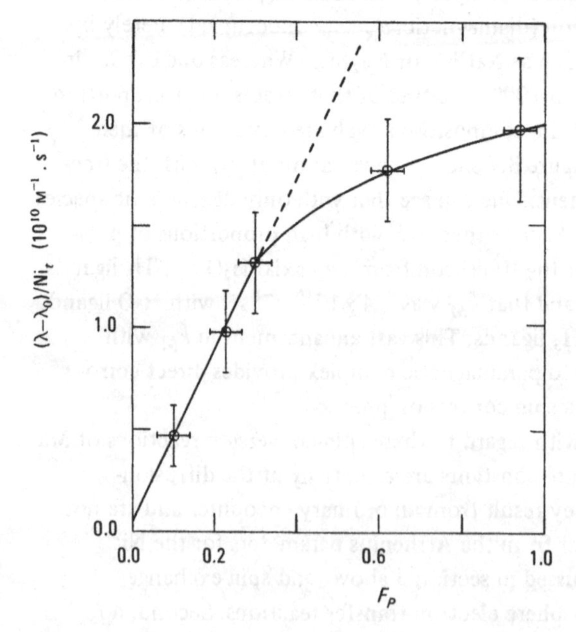

However, only three, with possible practical applications, will be given prominence here: the initiation reaction in the polymerization of vinyl monomers; the interaction of Mu with solutes inside micelles; and brief mention will be made of a few biologically-important systems.

Vinyl monomers

Table 8.6 presents the rate constant data for addition of Mu to five monomers (Stadlbauer *et al.*, 1981). The four acrylic compounds each react at close to the diffusion-limited rate, with small variations being attributable to differences in steric factors and electrophilic effects. There seems to be little doubt that these represent addition to the vinyl bond corresponding to initiation of polymerization - about which there is a dearth of data on H-atoms. This assignment as π-bond addition is supported by the k_M values in maleic and fumaric acids (1.1 and $1.4 \times 10^{10} \mathrm{M}^{-1} \mathrm{s}^{-1}$) which correspond closely to the monomers, whereas 2-hydroxyfumaric acid (with π-bonds saturated) reacts at only $k_M = 4.5 \times 10^7 \mathrm{M}^{-1} \mathrm{s}^{-1}$ (Percival, Roduner & Fischer, 1979). Unfortunately, k_H is known for only two of these monomers (see Table 8.6) and these give isotope ratios of 1.1 (no effect) and 2.8 (equal to the mean velocity ratio)!

The styrene situation is quite interesting. Its reaction rate with Mu is an order of magnitude slower than the diffusion-controlled limit, and significantly less than other aromatic compounds such as phenol and *p*-nitrophenol (see Table 8.2). Styrene also has a particularly small P_D yield (0.17) as with benzene (see Table 6.1), as if, in both cases, either epithermal Mu adds to the ring, or else π-delocalization strongly enhances thermalization of Mu* which then adds to the ring.

In order to try to sort this out, the structure of Mu-radicals formed in pure liquid styrene was studied by μSR, and compared with benzene and their mixtures (Stadlbauer *et al.*, 1981; Ng *et al.*, 1982). Some of the results are

Table 8.6. *Muonium addition reaction rate constants (k_M) with vinyl monomers in dilute aqueous solution at ~295 K*[a]

Monomer	k_M ($10^{10} \mathrm{M}^{-1} \mathrm{s}^{-1}$)	k_M/k_H[b]
Acrylamide (CH_2=$CHCONH_2$)	1.9	1.1 (± 0.3)
Acrylic acid (CH_2=$CHCO_2H$)	1.6	–
Acrylonitrile (CH_2=CHCN)	1.1	2.8 (± 0.4)
Methyl methacrylate (CH_2=$C(CH_3)CO_2CH_3$)	1.0	–
Styrene (C_6H_5CH=CH_2)	0.11	–

[a] Data taken from Stadlbauer *et al.*, 1981.
[b] k_H from Anbar *et al.*, 1975.

shown in Fourier transform in Figure 8.8. Firstly, there is only one pair of radical frequencies found in styrene. Therefore, regardless of where Mu initially adds to the styrene molecule, by the time of the μSR observation ($\sim 10^{-6}$ s) only one radical remains. This means either that Mu attacks only at one site, or that, from a distribution of initially formed radicals, only one survives intramolecular transformations or decay processes. A determination of the radical yields should be able to distinguish these alternatives; but such measurements are still inaccurate because of uncertainties about the response time of the μSR apparatus at such high frequencies. Secondly, the hyperfine coupling constant is 214 MHz for styrene, whereas it is 514 MHz for benzene (see Figure 8.8). By comparison with radicals derived from other aliphatic and aromatic compounds (see chapter 9) it is clear that the radical in styrene is exclusively that resulting from addition to the side-chain vinyl bond, not to the benzene ring.

Figure 8.8. Fourier transforms of the μSR spectra of Mu-containing free radicals produced in: (a) liquid styrene, and (b) benzene, both at 3500 G. In both cases the lowest frequency comes from diamagnetic muons. Hyperfine coupling constants can be equated with the sum of the two radical frequencies and are seen to be 214 MHz in styrene and 514 MHz in benzene.

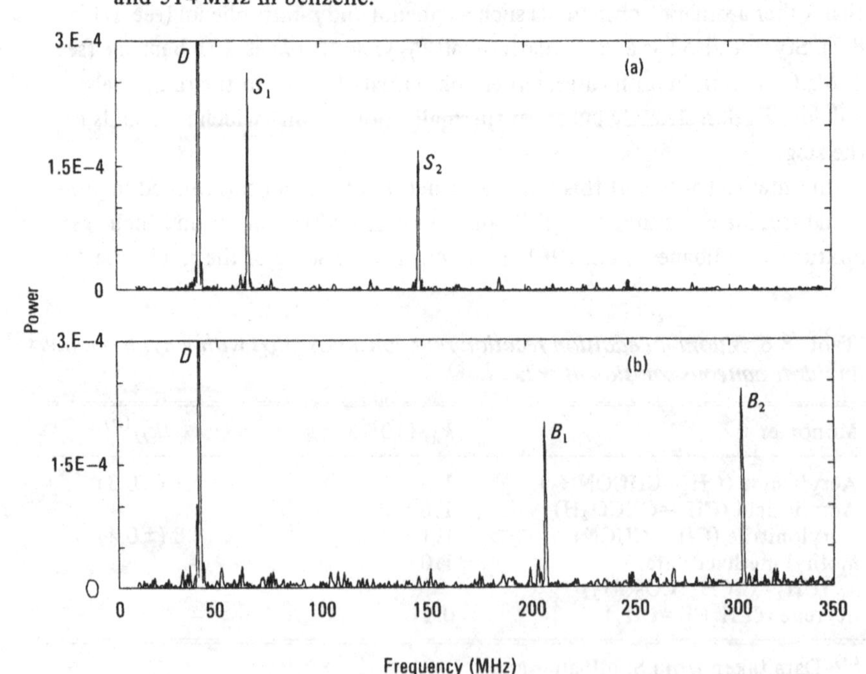

This does not explain why k_M is much smaller in styrene than in the other monomers - but it is consistent with the fact that polystyrene is produced with relatively low efficiency by free radical initiation compared to methylmethacrylate and other acrylic monomers (Mayo & Walling, 1950). There are other Mu reactions with π-bonds which are equally slow, or slower: for instance $k_M = 1.8 \times 10^9$ for ascorbic acid and $9 \times 10^7 \mathrm{M}^{-1} \mathrm{s}^{-1}$ for acetone (Percival *et al.*, 1979).

Micelles

One aspect of biophysical chemistry to which Mu studies may have something to contribute is in the elucidation of the effects of membrane and lipid bilayers on reactions requiring atom transport through these layers. Micelles have become one of the simplest types of chemical system used as models for this purpose as they are expected to mimic features of the real biological materials. A variety of physical techniques have consequently been applied to the study of micelles (Fendler & Fendler, 1975). One of these utilizes the ortho-positronium yield as determined from positron annihilation lifetime measurements (Jean & Ache, 1977). This signals the onset of the critical micelle concentration (CMC) - the small concentration range over which these surfactant molecules coalesce into the ordered micelle pattern indicated by Figure 8.9. As it happens, the muonium rate constants also change abruptly as the CMC is reached (Jean, Ng, Stadlbauer & Walker, 1981*a*).

In μSR experiments a solute was added at $\sim 10^{-4}$ M for reaction with the muonium atoms formed in the water. Solutes which are more soluble in hydrocarbons than in water were selected so they would all accumulate in the micelles. In practice, $\lambda - \lambda_0$ (equal to $k_M[S]$) was measured each time at a fixed solute (naphthalene) concentration as surfactant (sodium dodecyl sulphate) was progressively added to the solution. The rate constant was found to change sinusoidally, as in Figure 8.10, with the known CMC being close to the onset of the sharp increase in rate. This demonstrated that the μSR technique could be used to probe such properties of the micelle.

It also gave the interesting result that k_M increased rather than decreased at the CMC, showing that enclosure of naphthalene in a micelle actually enhances its reaction with Mu. It remains to be seen whether this enhancement stems from the fact that the actual reaction cross-section is greater in a hydrocarbon medium - even when ordered and viscous as in a micelle - or whether it reflects an increased encounter probability arising from the caging effect, or restricted diffusion, imposed by the micelle sheath once Mu is inside (Tachiya, 1980). In any event, it is clear that the muonium atom readily penetrates the micelle, so Mu is not seriously inhibited at the surface by the Gouy–Chapman or Stern layers, nor is it discouraged from permeating such ordered molecular aggregations (see Figure 8.9).

Systems of biological interest

One facet of muonium chemistry with potential use in studying 'the biochemistry of life' lies in its reactions with transition metal ions and the effect of their ligands on rates. Some k_M values are bound to be of significance here, because Mu is one of the strongest available reducing agents. Furthermore, it has the unique property of registering spin-conversion interactions, so it can

Figure 8.9. A two-dimensional representation of the various regions of a spherical micelle formed from an anionic surfactant in water. The counterions (X), head groups (dotted circles), and hydrocarbon chains (zig-zag lines) are indicated schematically to denote their relative locations but not their number, distribution or configuration. (Figure from Fendler & Fendler, 1975.)

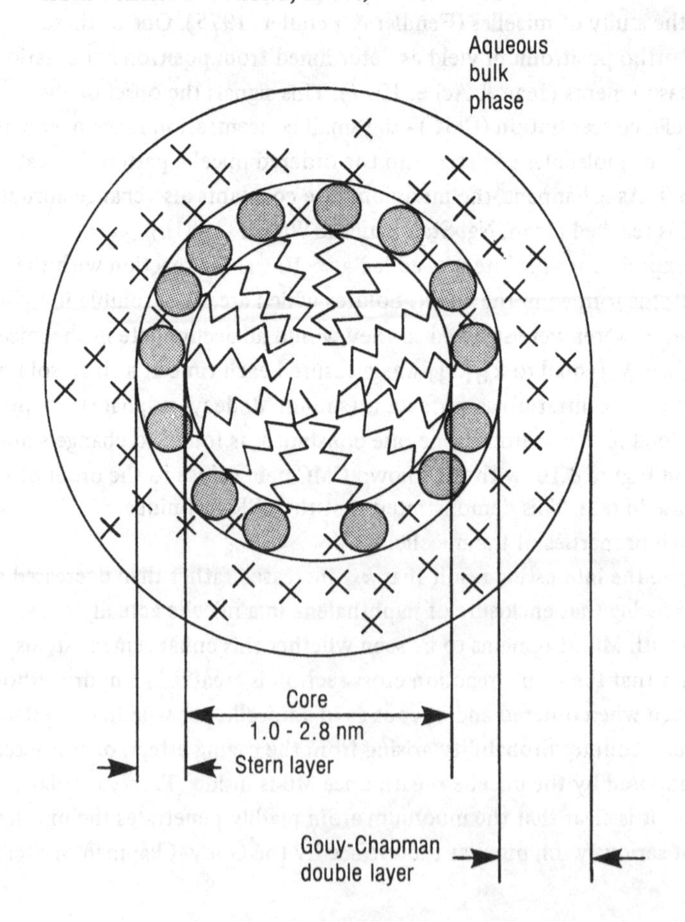

disclose paramagnetism and therefore distinguish different spin states in some cases.

It was shown, for instance, that muonium reactivity is sensitive to the presence or absence of Fe(III) in a porphyrin. Thus Jean *et al.* (1979*a*) found the reaction with hemin to have a rate constant of $2.3 \times 10^9 \text{M}^{-1}\text{s}^{-1}$ compared to $0.62 \times 10^9 \text{M}^{-1}\text{s}^{-1}$ for protoporphyrin. This latter compound has the same composition except that the Fe(III) of hemin is replaced by two protons. Although an order of magnitude smaller than reaction with $Fe(CN)_6^{3-}$ (see Table 8.7) and the diffusion-controlled limit, this reaction rate towards hemin is probably consistent with the shielding effect of the porphyrin sheath - since the Fe can only be attacked through restricted axial approaches.

In a similar vein were studies of solutes present as inclusion compounds in cyclodextrins in dilute aqueous solutions (Jean *et al.*, 1981*b*). The α and β cyclodextrins are doughnut-shaped molecules comprising six (for α) or seven (for β) glucose units linked in a ring with inner cavities of 1-1.2 nm diameter. These molecules provide two different environments, an internal hydrophobic surface

Figure 8.10. Values of the rate constant for the reaction of muonium with naphthalene at 10^{-4} M, plotted against the concentration of sodium dodecyl sulphate (NaLS) monomer units added to the solution. Error bars given are two statistical standard deviations. The known critical micelle concentration (CMC) of NaLS is indicated by the arrow. (Figure from Jean *et al.*, 1981*a*.)

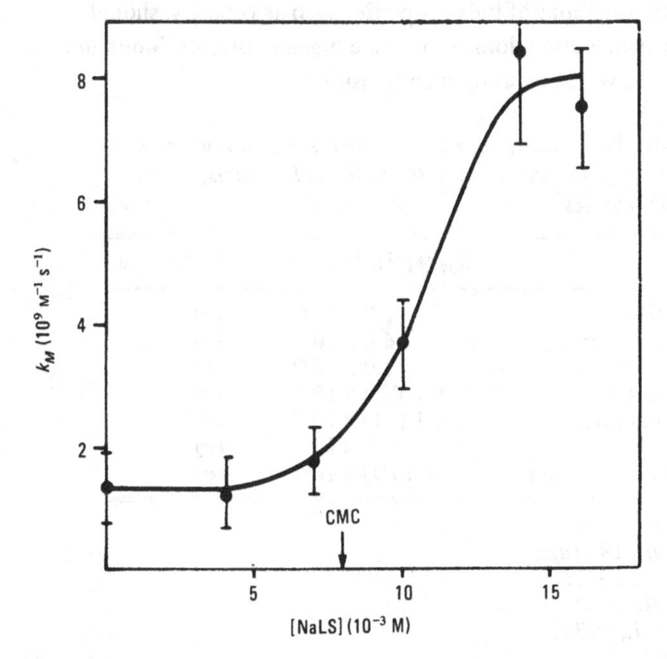

in the central cavity in which guest molecules may be occluded, and an outer hydrophilic surface arising from the OH groups. With I_2 as the inclusion-compound, the reaction rate constant of Mu was found to be essentially the same as that with free I_2 in dilute aqueous solution (see Table 8.7). Apparently the presence of the sugar sheath enclosing the solute barely affects the reaction rate: k_M remains close to the diffusion-controlled limit despite the sheath to penetrate and the bonding of I_2 to the cyclodextrin to overcome (albeit mainly weak van der Waals forces). Porphyrin ligands are evidently much more inhibiting to Mu than cyclodextrin or micelle sheaths, or octahedral cyano ligands.

Bucci *et al.* (1979) have reported briefly on rate data for the reactions of Mu with DNA and constituent nucleotides, including those of thymine, cytosine, adenine, guanine and uracil. In general k_M was in the range 4 to $9 \times 10^9\ M^{-1}s^{-1}$ and there were isotope effects of >10 favouring Mu over the corresponding H reaction. Thymine and adenine nucleotides reacted faster than the others; but with DNA itself there were the usual uncertainties in establishing the concentration and the true organizational state of the strands in solution.

Finally, it should be recalled that all the above refers to dilute solutions in water. Some reaction-rate studies are being pursued in alcohol and alkane media, and they seem to give rate constants higher than in water. One would expect this for the diffusion-limited rates because of the lower viscosities. Furthermore, the transitions from activation-controlled, to diffusion-controlled, to tunnelling-controlled, which are functions of bulk properties such as viscosity, should register in different temperature domains in these organic solvents. Muonium is an ideal reagent with which to study such transitions.

Table 8.7. *Muonium reaction rates in aqueous solution at ~295 K for a few species of interest as model biological systems, or for comparison purposes*

Compound	$k_M\ (M^{-1}s^{-1})$	Reference
Hemin (Fe III)	2.7×10^9	(*a*)
Protoporphyrin (no Fe)	6.0×10^8	(*a*)
$Fe(CN)_6^{3-}$	2.0×10^{10}	(*b*)
I_2 in cyclodextrins	$1.3\ (\pm 0.3) \times 10^{10}$	(*c*)
I_2 in NaOS micelles	$4.1\ (\pm 1) \times 10^{10}$	(*d*)
I_2	1.7×10^{10}	(*c*)
DNA and its nucleotides	$(4 \text{ to } 9) \times 10^9$	(*e*)

(*a*) Jean *et al.*, 1980.
(*b*) Jean *et al.*, 1978*a*.
(*c*) Jean *et al.*, 1981*b*.
(*d*) Jean *et al.*, 1981*a*.
(*e*) Bucci *et al.*, 1979.

References to chapter 8

Ache, H. J. (1979). Positronium chemistry: present and future directions. *Advances in Chemistry Series*, **175**, 1-47.

Allen, A. O. (1976). Drift mobilities and conduction band energies of excess electrons in dielectric liquids. *National Standard Reference Data Systems - National Bureau of Standards*, No. 58.

Anbar, M., Bambenek, M. & Ross, A. B. (1973). Selected specific rates of reactions of transients from water in aqueous solution. 1. Hydrated electrons. *National Standard Reference Data Systems - National Bureau of Standards*, No. 43.

Anbar, M., Farhataziz, & Ross, A. B. (1975). Selected specific rates of reactions of transient species from water in aqueous solution. 2. Hydrogen atoms. *National Standards Reference Data Systems - National Bureau of Standards*, No. 51.

Baxendale, J. H. (1964). Redox potential and hydration energy of the hydrated electron. *Radiation Research*, Supplement **4**, 139-40.

Brewer, J. H., Crowe, K. M., Johnson, R. F., Schenck, A. & Williams, R. W. (1971). Fast depolarization of positive muons in solution - the chemistry of atomic muonium. *Physical Review Letters*, **27**, 297-300.

Bucci, C., Guidi, G., Manfredi, M., Tedeschi, R., Crippa, P. R., de Munari, G. M., Vecli, A. & Podini, P. (1979). Interaction of muonium with molecules of biological interest in water solution. *Hyperfine Interactions*, **6**, 425-9.

Dodelet, J.-P. & Freeman, G. R. (1972). Mobilities and ranges of electrons in liquids. Effect of molecular structure in C_5-C_{12} alkanes. *Canadian Journal of Chemistry*, **50**, 2667-79.

Fendler, J. H. & Fendler, E. J. (1975). *Catalysis in Micellar and Macromolecular Systems*. New York: Academic Press.

Firsov, V. G. & Byakov, V. M. (1965). Chemical reactions involving muonium. A method for determining the absolute rate constants and other reaction parameters. *Soviet Physics JETP*, **20**, 719-25.

Goldanskii, V. I. (1968). Physical chemistry of the positron and positronium. *Atomic Energy Reviews*, **6**, 3-148.

Hammett, L. P. (1970). *Physical Organic Chemistry*, 2nd edn, pp. 201-6. New York: McGraw-Hill Book Co.

Hart, E. J. & Anbar, M. (1970). *The Hydrated Electron*. New York: Wiley-Interscience.

Ivanter, I. G. & Smilga, V. P. (1968). Theory of the muonium mechanism of depolarization of μ^+ mesons in media. *Soviet Physics JETP*, **29**, 301-6.

Jean, Y. C. & Ache, H. J. (1977). Positronium reactions in micellar systems. *Journal of the American Chemical Society*, **99**, 7504-9.

Jean, Y. C., Brewer, J. H., Fleming, D. G., Garner, D. M., Mikula, R. J., Vaz, L. C. & Walker, D. C. (1978*a*). Reactivity of muonium atoms in aqueous solution. *Chemical Physics Letters*, **57**, 293-7.

Jean, Y. C., Brewer, J. H., Fleming, D. G. & Walker, D. C. (1978*b*). Spin-conversion of muonium by interaction with paramagnetic ions. *Chemical Physics Letters*, **60**, 125-9.

Jean, Y. C., Brewer, J. H., Fleming, D. G., Garner, D. M. & Walker, D. C. (1979*a*). Determination of the effective charge on muonium during its reactions in aqueous solution. *Hyperfine Interactions*, **6**, 409-12.

Jean, Y. C., Fleming, D. G., Ng, B. W. & Walker, D. C. (1979*b*). Reaction of muonium with O_2 in aqueous solution. *Chemical Physics Letters*, **66**, 187-90.

Jean, Y. C., Ng, B. W. & Walker, D. C. (1980). Chemical reactions between muonium and porphyrins. *Chemical Physics Letters*, **75**, 561-4.

Jean, Y. C., Ng, B. W., Stadlbauer, J. M. & Walker, D. C. (1981*a*). Muonium reactions in micelles. *Journal of Chemical Physics*, **75**, 2879-83.

Jean, Y. C., Ng, B. W., Ito, Y., Nguyen, T. Q. & Walker, D. C. (1981*b*). MSR applications to muonium reactivity in cyclodextrins. *Hyperfine Interactions*, **8**, 351-4.

Lazzarini, A. L. F. & Lazzarini, E. (1980). Influence of the isotopic composition of the water on the rate constant of the reaction between *o*-Ps and K[Co(EDTA)]. *Journal of Inorganic and Nuclear Chemistry*, **42**, 953–6.
Mayo, F. R. & Walling, C. (1950). Copolymerization. *Chemical Reviews*, **46**, 191–287.
Mobley, R. M., Amato, J. J., Hughes, V. W., Rothberg, J. E. & Thompson, P. A. (1967). Muonium chemistry II. *Journal of Chemical Physics*, **47**, 3074–5.
Molin, Yu. N., Salikhov, K. M. & Zamaraev, K. I. (1980). *Spin Exchange*. Berlin: Springer-Verlag.
Nagamine, K., Nishiyama, K., Imazato, J., Nakayama, H., Yoshida, M., Sakai, Y., Sato, H. & Tominaga, T. (1982). Long-lived muonium in water revealed by pulsed muons. *Chemical Physics Letters*, **87**, 186–91.
Ng, B. W., Jean, Y. C., Ito, Y., Suzuki, T., Brewer, J. H., Fleming, D. G. & Walker, D. C. (1981). Diffusion- and activation-controlled reactions of muonium in aqueous solutions. *Journal of Physical Chemistry*, **85**, 454–8.
Ng, B. W. *et al.* (1982). To be published.
Nichols, A. L., Wild, R. E., Bartal, L. J. & Ache, H. J. (1974). Positronium atom reactions in aqueous solutions of Fe III complexes. *Applied Physics*, **4**, 37–40.
Percival, P. W., Fisher, H., Camani, M., Gygax, F. N., Ruegg, W., Schenck, A., Schilling, H. & Graf, H. (1976). The detection of muonium in water. *Chemical Physics Letters*, **39**, 333–5.
Percival, P. W., Roduner, E., Fischer, H., Camani, M., Gygax, F. N. & Schenck, A. (1977). Bimolecular rate constants for reactions of muonium in aqueous solution. *Chemical Physics Letters*, **47**, 11–14.
Percival, P. W., Roduner, E. & Fischer, H. (1978). Radiolysis effects in muonium chemistry. *Chemical Physics*, **32**, 353–67.
Percival, P. W., Roduner, E. & Fischer, H. (1979). Radiation chemistry and reaction kinetics of muonium in liquids. *Advances in Chemistry Series*, **175**, 335–5.
Pilling, M. J. (1975). *Reaction Kinetics*. London: Oxford University Press.
Roduner, E. (1979). *On the Liquid Phase Chemistry of the Light Hydrogen Isotope Muonium*. Ph.D. Thesis, University of Zurich, pp. 1–97.
Roduner, E. & Fischer, H. (1979). The importance of tunneling in chemical reactions of muonium. *Hyperfine Interactions*, **6**, 413–17.
Stadlbauer, J. M., Ng, B. W., Walker, D. C., Jean, Y. C. & Ito, Y. (1981). Muonium addition to vinyl monomers. *Canadian Journal of Chemistry*, **59**, 3261–6.
Stadlbauer, J. M., Ng, B. W., Jean, Y. C. & Walker, D. C. (1982). Spin conversion reaction of muonium with nickel cyclam. *Journal of the American Chemical Society* (in press).
Tachiya, M. (1980). Diffusion-controlled reaction in a micelle. *Chemical Physics Letters*, **69**, 605–7.
Walker, D. C. (1981). Muonium: a light isotope of hydrogen. *Journal of Physical Chemistry*, **85**, 3960–71.

9

FREE RADICALS CONTAINING MUONS

9.1 Preamble

Free radicals are recognized to be important reactive intermediates in many branches of chemistry and biology, so all aspects of their formation, structure and reactivity are of widespread interest. Hitherto, most free radical studies have relied on magnetic resonance techniques utilizing the absorption of electromagnetic radiation; but in the muon spin rotation method, one observes the evolution of the muon's spin polarization - as modified by its hyperfine interaction with the unpaired electron in a free radical containing the muon. Already the ability of μSR to study μ^+-containing free radicals can be regarded as one of its most valuable assets.

The radicals involved are generally those resulting from formal addition of Mu to a double bond, so the muon is attached to an atom at least one removed from the atom carrying the majority of the unpaired electron density. But because of its unique nuclear moment, the muon probes that electron density differently to protons around it in equivalent positions. There are certainly isotope effects of considerable interest and use - both in the hyperfine coupling constants and in reactivity of Mu compared with H - but the muon is primarily acting here as a distinguishable nuclear probe. Fortunately, the observational timescale of μSR coincides with typical free radical lifetimes in media of prime interest. This allows both structure and kinetics to be studied. In fact one can even draw inference about reaction rates back into the subnanosecond timescale, because of the requirement in μSR of initial phase coherence in the precession of the species observed at times $>10^{-7}$ s.

μSR free radical chemistry has surfaced in the last four years. The basic technique, and its variants, were pioneered and developed by Roduner and Fischer, and their collaborators, as the references cited in this chapter will attest. The occurrence of Mu-radicals was predicted by Brodskii (1963) and followed by a period in which they were implicated as intermediates by the absence of inflections in P_{res} titration curves (Brewer *et al.*, 1974). But the problems associated with their direct observation – demonstrated by the efforts of Bucci *et al.* (1978) in trying to analyse the enormously complex spectrum found in an aqueous solution of thymine at low magnetic fields – was neatly solved by Roduner *et al.* (1978) by decoupling the nuclear spins in the Paschen-Back high-field limit, as used in ENDOR. In previous theoretical considerations (Percival & Fischer, 1976) the radicals had been regarded as two spin-1/2 variants of Mu with reduced hyperfine coupling, because of the truncated Hamiltonian used; but when terms were added for the other non-zero nuclear spins (Roduner & Fischer, 1979, 1981; Roduner, 1979) it became clear that in low magnetic fields the muon-spin polarization in the radicals would be dispersed over many precession frequencies. So much so, that most of their intensities would be too small to observe or characterize. Even with only one proton in addition to μ^+ and e^-, there are 15 frequencies involved at 20 G, as shown in Figures 9.1 and 9.2. However, at high enough fields (>1 kG) the electron Larmor frequency exceeds the muon and proton hyperfine frequencies,

Figure 9.1. Calculated μSR spectra (amplitudes versus frequencies) for the μ^+-e^--p^+ system in different (low) magnetic fields. (A_p = 56 MHz and A_μ = 178 MHz). (Figure taken from Roduner & Fischer, 1981.)

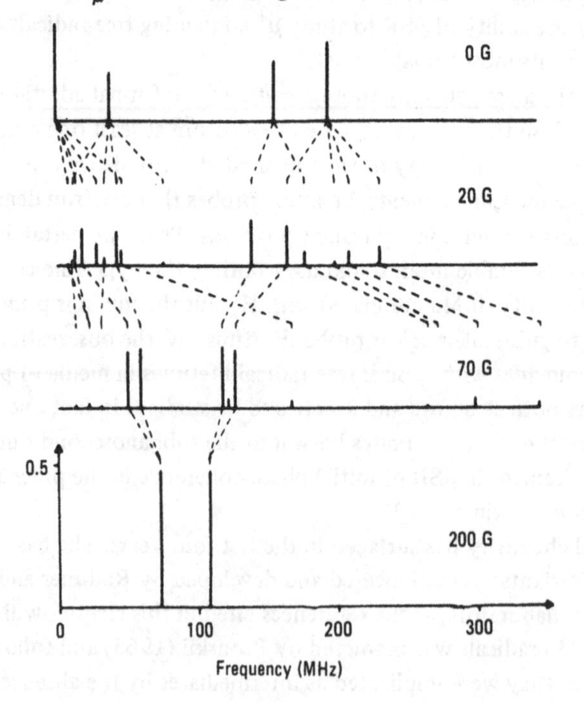

so that the radical frequencies degenerate to two, with the whole polarization distributed equally between them (see Figures 9.1 and 9.2).

In this high-field limit, the isotropic muon-electron hyperfine coupling constant (A_μ here, by analogy with symbols used in ESR, and not to be confused with A_μ as previously used for the diamagnetic muon asymmetry - A_D here) is related to the precession frequencies (R in Figure 3.5) by Eqs. [9.1] and [9.2],

$$\nu_{12} = \tfrac{1}{2}A_\mu - \nu_D \text{ and } \nu_{43} = -\tfrac{1}{2}A_\mu - \nu_D \qquad [9.1]$$

$$A_\mu = \nu_{12} - \nu_{43} \quad \text{or} \quad A_\mu = |\nu_{12}| + |\nu_{43}| \qquad [9.2]$$

where ν_D is the free muon precession frequency and ν_{12} and ν_{43} have opposite signs at fields $\lesssim 7$ kG. In practice, then, one merely sums the two radical frequencies to obtain A_μ in MHz.

Most μSR free radical spectra are taken with the same apparatus as that used in free Mu and P_D studies (see Figure 9.3 for yet another example of μSR arrangements). The only special requirement is that magnetic fields up to 5 kG

Figure 9.2. Calculated field dependence of the μSR frequencies for the μ^+-e^--p^+ system in transverse fields. μ^+ denotes the diamagnetic muon precession frequencies. Broken lines indicate low intensities. (Figure taken from Roduner & Fischer, 1981.)

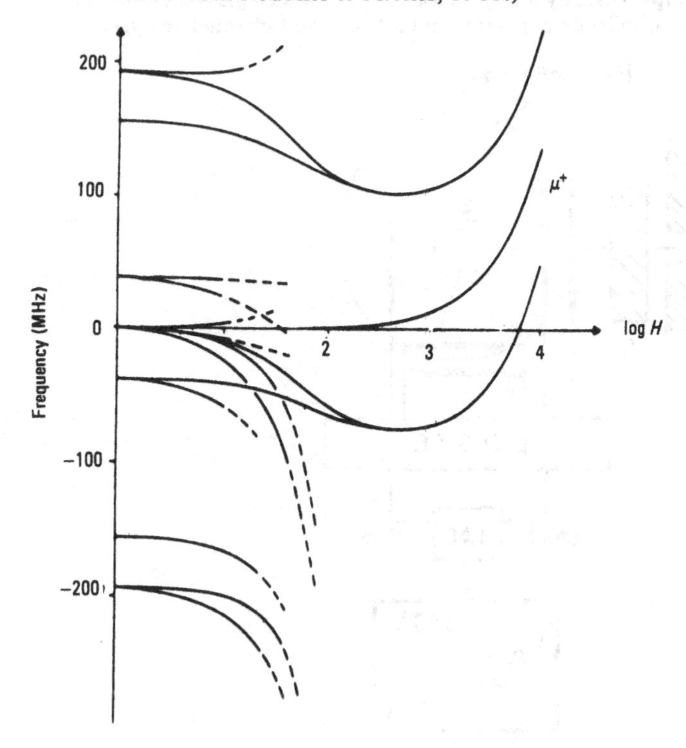

or so are required which, in transverse mode with longitudinally polarized muons, seriously bend low energy (surface) muon beams. Samples need to be sufficiently well deoxygenated so that reactions to peroxy-radicals do not occur in $<10^{-5}$ s. Radicals have been observed in most pure organic compounds containing π-bonds. They may be studied as pure (neat) liquids or in concentrated solution in an inert solvent in which Mu is formed. They can also be studied as solids, in single crystals, powders or amorphous states; but anisotropies and line-broadening effects creep in. The only chemical limitations are these: that Mu (or whatever is the radical precursor) react sufficiently quickly ($<10^{-9}$ s) so that any spin-dephasing is small; and second that the radical be sufficiently long-lived ($>10^{-6}$ s) for the μSR observation. μSR spectra for radical studies are usually shown after Fourier transformation (Brewer *et al.*, 1982) since presentation in frequency-space rather than time-space makes A_μ characterization easier.

Of the radicals reported so far, all have the μ^+ present as a muonium atom, bonded, like an H atom, to C or O through an ordinary σ-bond. Thus they are

Figure 9.3. Schematic of μSR equipment for use with high energy (backward) muons. a, b, c are the muon (start) counters, d, e, the positron (stop) counters, g the beam moderator, f the collimator, and S is the sample (usually a degassed liquid in a sealed glass sample). TAC is a time-to-amplitude converter and MCA a multichannel analyser.

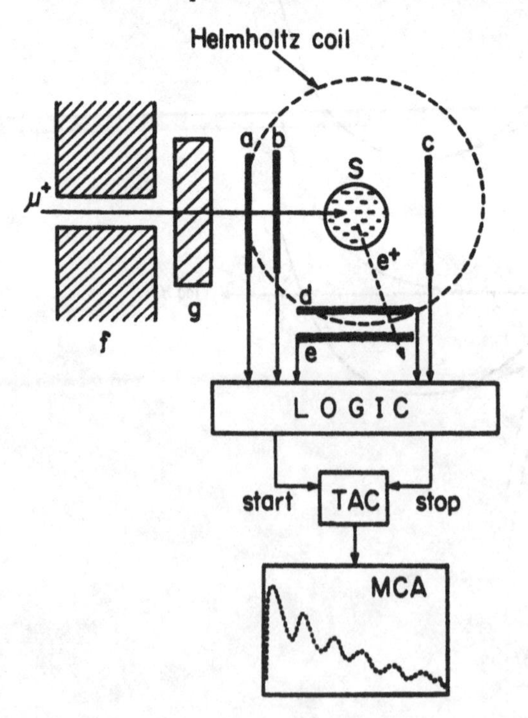

all describable as Mu-radicals, though the Mu atom may be bonded more than one atom removed from the radical centre. However, the possibility should not be ruled out of observing other types of muon-containing free radical species such as radical cations or anions, biradicals, or paramagnetic complexes.

As there are several aspects to these free radical studies, this chapter is divided into sections dealing separately with: coupling constants; isotope effects; selectivity in Mu-radical formation; mechanism of formation and yields; reactions of radicals; information available at low magnetic fields; and anisotropic effects in oriented single crystals.

9.2 Isotropic muon-electron hyperfine coupling constants (A_μ)

All Fourier transform μSR spectra show the peak at 0.0136 MHz G^{-1} due to the presence of some muons in diamagnetic states. This represents only 15% of the total muon polarization in benzene, but some 45% in mono-alkenes. It serves as a useful reference in both frequency and amplitude. On the other hand, no free Mu is observable at 1.39 MHz G^{-1} in any unsaturated compound, consistent with the fact that Mu attacks all π-bonded material, as it does in dilute aqueous solution with k_M values ranging from 10^8 to $3 \times 10^{10}\,M^{-1}\,s^{-1}$ (see Appendix). In its place are two peaks in high fields corresponding to the radical formed. The first such direct observation of a Mu-radical is shown for pure liquid 2,3-dimethyl-2-butene in Figure 3.8 (Roduner *et al.*, 1978) corresponding to addition as in Eq. [9.3].

$$Mu + (CH_3)_2C{=}C(CH_3)_2 \rightarrow (CH_3)_2\dot{C}{-}C(CH_3)_2(Mu) \qquad [9.3]$$

Table 9.1. *Hyperfine coupling constants (A_μ) for the Mu-radical formed from a selection of π-bonded molecules, and the isotope effect relative to H (after compensation for the nuclear moment ratio, 3.18)*

Molecule	Assigned radical	A_μ (MHz)	$A_\mu/A_p \times 3.18$	Reference
acetone	$(CH_3)_2\dot{C}O(Mu)$	26.0	9.1	(*a*)
2.3-dimethyl-2-butene	$(CH_3)_2\dot{C}C(CH_3)_2(Mu)$	161.0	1.68	(*a*)
butadiene	$CH{=}CH\dot{C}HCH(Mu)$	188.3	1.29	(*b*)
hexafluorobenzene	$C_6F_6(Mu)$	200.9	–	(*c*)
styrene	$C_6H_5\dot{C}HCH_2(Mu)$	213.5	1.34	(*b*), (*d*)
acrylonitrile	$NCCH\dot{C}H_2(Mu)$	266	1.30	(*b*), (*d*)
2-methylpropene	$(CH_3)_2\dot{C}CH_2(Mu)$	291.6	1.44	(*b*)
1-pentene	$C_3H_7\dot{C}HCH_2(Mu)$	306.9	1.39	(*b*)
pentafluorobenzene	$C_6HF_5(Mu)$	453.4	–	(*c*)
benzene	$C_6H_6(Mu)$	514.6	1.21	(*a*)
benzene-d$_6$	$C_6D_6(Mu)$	520.1	1.20	(*a*)

(*a*) Roduner, 1979: (*b*) Roduner *et al.*, 1982*a*.
(*c*) Roduner *et al.*, 1982*b*. (*d*) Stadlbauer *et al.*, 1981.

It consists of the free muon peak at ν_D and two peaks whose mean value ($A_\mu/2$) is seen to be field-independent and whose individual values are displaced from the mean by $\pm\nu_D$ at all fields, in accord with Eqs. [9.1] and [9.2].

Some 70 unsaturated organic compounds which give Mu-radicals have now been studied in this way. Their A_μ values are collected in the Appendix for quick reference and a very small selection is given in Table 9.1. Not only can comparison with the coupling constants of their H-analogues be used in characterization, but these now constitute a sufficiently large body of information that their own interrelationships and self-consistent patterns serve as an independent source of identification.

Mu-radical coupling constants are found to cover a vast range: from 26 MHz in the Mu-acetone adduct to 520 MHz for the cyclohexadienyl-d_6 radical in C_6D_6. Both large and small effects appear. One of the most striking is the transition from C_6HF_5 (453 MHz) to C_6F_6 (200.9 MHz). At the other extreme it is possible to distinguish most isomers and identify different directional trends on aromatic rings. Take as an example the addition of Mu to $C_6H_5CF_3$. Four Mu-radicals are found and identified through their A_μ values: with Mu adding at the *meta* position, $A_\mu = 510.5$ MHz; at the *para*, 508.7 MHz; at the *ortho*, 500.3 MHz; and at the *ipso* position, 474 MHz (Roduner, Brinkman & Lowrier, 1982*b*). The yields of each are different, so the relative selectivity-factors can also be determined. Sometimes the selectivity is totally effective. Thus in styrene only one of six possible radicals is observed (see Figure 8.8), and from its A_μ value (see Table 9.1) it is evident that Mu binds exclusively to the vinyl side chain (Stadlbauer *et al*., 1981).

Figure 9.4. Representation of the dihedral angle θ. C_α is the far carbon containing the half-filled p orbital at the free-radical centre. C_β is the nearer atom containing 2H and Mu.

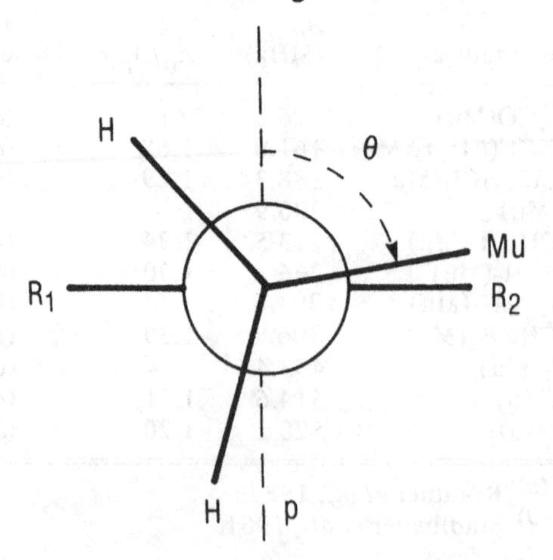

9.3 Isotope effects

Muon and proton hyperfine coupling constants (A_μ and A_p, respectively) should be in the ratio 3.18 due to their respective nuclear magnetic moments (see Table 3.1). Any further differences represent 'dynamic' isotopic effects. It transpires that for all cases for which there are unequivocal data and assignments, A_μ is greater than $3.18A_p$. The magnitude of these residual isotope effects are given in the fourth column in Table 9.1 for that selection of compounds but are now known for many of the compounds in the Appendix. Favouring of Mu compared to H (as a ratio) is smallest in the aromatics (1.1-1.3), intermediate for the aliphatics (1.3-1.7), and largest, so far, for addition to the O atom of a carbonyl (9.1 for acetone).

These isotope effects have been explained by reference to the radicals' internal dynamics and degrees of freedom (Roduner, 1979; Roduner *et al.*, 1982*a*). For the cyclohexadienyl radicals from aromatic compounds, the differences between Mu and H are attributed to more pronounced out-of-plane deformation by the >CH(Mu) group due to the larger amplitude of the Mu vibrations. For the radicals formed by addition to a carbonyl, the isotope effect is attributed to a greatly reduced moment of inertia in the O-Mu com-

Figure 9.5. Temperature dependence of the hyperfine coupling constants of $XCH_2\dot{C}(CH_3)_2$ radicals where X is Mu (top), ^1H (middle) and ^2H (bottom), normalized to the ^1H values in accordance with the ratio of nuclear moments of Mu and ^2H relative to ^1H. (Figure taken from Roduner *et al.*, 1982*a*.)

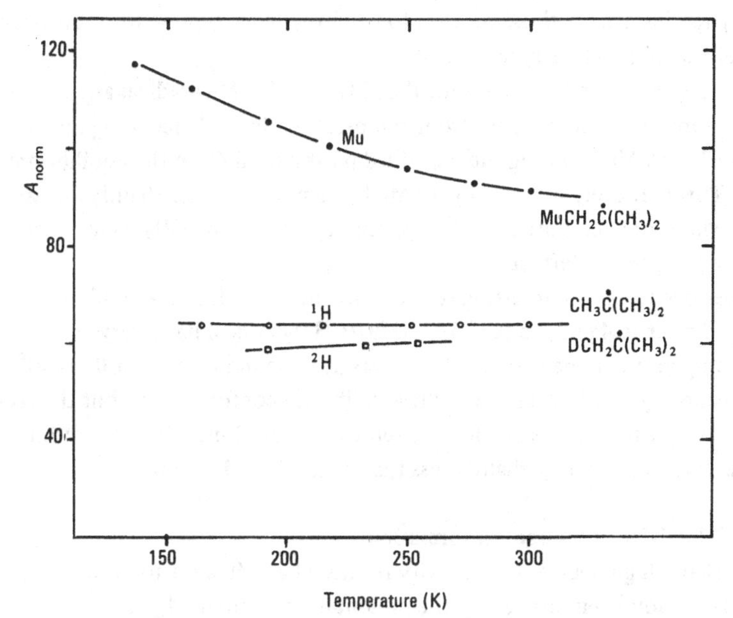

pared to O–H group rotating about the C–O axis of the radical. Finally, for the aliphatic compounds, the isotopic difference is ascribed to incomplete rotational averaging, with unequal contributions from all conformers arising from the fact that Mu experiences more steric hindrance when eclipsed, than does H, because of its higher zero-point vibrational motion.

This latter explanation follows that formulated by Fessenden (1964) to account for differences between ^1H and ^2H radicals. It can most easily be illustrated using terminal CH_3 or $CH_2(Mu)$ groups in the β-position to the radical centre (Roduner, 1979). The instantaneous value of a coupling constant depends on the dihedral angle (θ), which is the angle between the Mu bond and the axis of symmetry of the unpaired electron *p*-orbital on C, as illustrated in Figure 9.4 using a Newman projection. '*A*' values obey Eq. [9.4],

$$\langle A\rangle = A_0 + B\langle\cos^2\theta\rangle \qquad [9.4]$$

where B describes the angular dependence of the spin density at C_β (Heller & McConnell, 1960). B is related to the magnetic moment and is generally much larger than the constant A_0 arising from spin polarization contributions. The observed A is a statistically weighted average of the conformational positions, so it depends on the reduced moment of inertia and on the barrier to internal rotation about the C_α-C_β axis. When there is a significant barrier, A is temperature-dependent with an expectation value in the high-temperature limit (free rotation) equal to $\frac{1}{2}A_0 + B$, since $\cos^2\theta = \frac{1}{2}(1 + \cos 2\theta)$. As the temperature is lowered, A should increase for those radicals with equilibrium conformations at $\theta_0 < 45°$, and decrease for $45° < \theta_0 < 90°$. In the case of CH_3, its three-fold symmetry renders A_p for those protons to be virtually at the free-rotation value and therefore temperature-independent.

Results are shown in Figure 9.5 for the $(CH_3)_2\dot{C}CH_2(Mu)$ radical and its ^1H and ^2H isotopic equivalents. The strong temperature dependence in A_μ indicates that $\theta_0 \sim 0°$ with Mu eclipsing the half-filled p-orbital on C_α in the equilibrium position. This is entirely in accord with Mu being much bulkier than H due to its higher vibrational amplitudes. The opposite applies to the CH_2D group, hence its positive temperature effect.

Another particularly interesting result along the same lines was Roduner's (1979) finding that A_μ of the $(CH_3)_2\dot{C}C(Mu)(CH_3)_2$ radical has a very small positive temperature-dependence. That it was positive indicated that $\theta_0 = 90°$, with Mu eclipsing a methyl group in the equilibrium conformation; but the fact that it was very small suggested that the ferocious vibrations of the Mu atom create steric hindrance only slightly less than that of a CH_3 group.

9.4 Selectivity in Mu-radical formation

That a high degree of selectivity results when Mu adds to double bonds, has already been implied by the presentation of single A_μ values in

Table 9.1. Mu adds almost exclusively to the O in acetone, to the terminal bond in 2-methyl propene, to the least-substituted C in pentene-1, to give only allylic radicals in butadiene, and just a side-chain radical in styrene. This closely follows the selectivity pattern shown by H (Fischer & Hellwege, 1977–9).

For non-symmetric dienes only two of the four possible radicals are found (Roduner *et al.*, 1982*a*), those in which π-delocalization extends over three adjacent C atoms. A typical two-radical spectrum in a single-component medium is shown as Figure 9.6. Pairing the four lines is readily achieved by varying the magnetic field, because the mean frequency of each pair remains unchanged. When alternative allylic radicals are possible, as in 2.4-dimethyl-1.3-pentadiene, only the one with the lower coupling constant occurs. Methyl and fluoro substituents on benzene are found to be *ortho*-directing towards Mu addition, and C–H sites are favoured over *ipso* positions by at least a factor of three (Roduner *et al.*, 1982*b*). Furthermore, among the radicals formed in toluene, the ratio of *ortho*/*para*/*meta* in Mu-addition is the same as that found for thermal tritium addition, and hot (recoil) addition. The implication is, that whatever factors may change the kinetic isotope effect, they are basically position-independent.

Figure 9.6. A representative Fourier transform μSR spectrum (power versus frequency), in 2,4-hexadiene at ~295 K. D marks the diamagnetic muon frequency; Cy is an experimental artifact being the cyclotron frequency; R_1 and R_2 are the two pairs of frequencies representing two radicals formed from this molecule in quite different yields. (Figure taken from Roduner *et al.*, 1982*a*.)

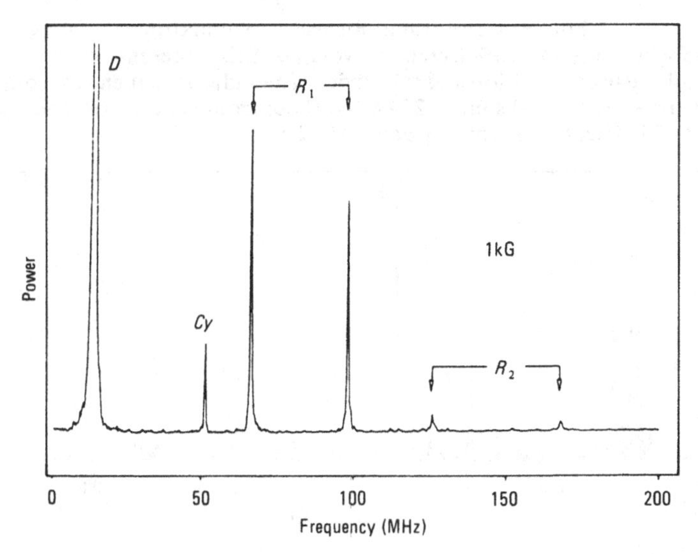

There is also *inter*molecular selectivity shown. When mixtures of benzene and styrene were examined they showed the same two Mu-radicals (A_μ values of 214 and 514 MHz) as do the separate liquids; but the ratio of their intensities was not proportional to the volume ratios of the two liquids (Ng *et al.*, 1982). Instead, the two radicals occurred with approximately the same intensity (these liquids have almost equal P_D values) when the mixture consisted of 85% benzene–15% styrene, as shown in Figure 9.7. Rather than following indiscriminate reactivity towards whichever molecule it happens to be in contact with, the Mu-radical precursor shows competitive kinetics – with inverse-yield versus inverse-concentration plots being essentially linear. It reacts with styrene six times faster than with benzene – despite the fact that either intramolecular rearrangement or initial site selectivity is essential with styrene, since only the radical formed by vinyl addition is observed.

One can also draw the following conclusion from these results. Once the Mu-radicals of either benzene or styrene are formed there is no further intermolecular conversion on the μSR timescale, because both radicals are observed in some proportion in all mixtures. One can also infer that cyclohexadiene radicals do not add to styrene to initiate its polymerization on this time-scale.

Selectivity can appear in the primary radical-forming step, and be basically kinetic in origin, or it may arise from thermodynamic factors which cause rearrangement to the most stable radical structure. Among kinetic factors would be steric hindrance, relative size or number of equivalent π-bond sites, and different activation barriers created by electromeric effects. Perhaps kinetic factors dominate intermolecular selectivity, while thermodynamic factors control intramolecular selectivity.

Figure 9.7. Fourier transform μSR spectra in a mixture consisting of 85% benzene and 15% styrene by volume. This represents intermolecular competition and selectivity. The radical frequencies from styrene (S_1 and S_2) sum to 214 MHz, those from benzene (B_1 and B_2) to 514 MHz. (Data from Ng *et al.*, 1982.)

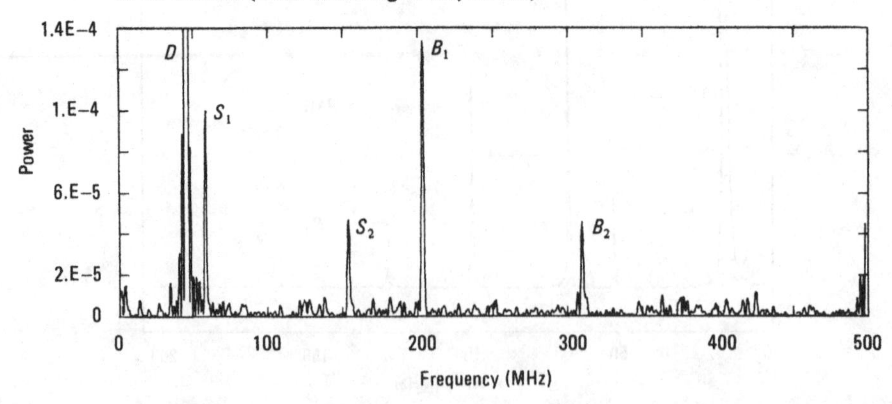

9.5 Mechanisms of Mu-radical formation, and their yields

There are several possible mechanisms by which Mu-radicals could be formed with the above selectivity, including the following: (i) Addition of thermalized muonium atoms to the π-bonded substrate under competitive kinetic conditions, involving activation barriers sufficiently high that almost complete selectivity for the initial reaction site can result. Such selective reactions would have to occur rapidly enough to avoid depolarization of ^{S}Mu or dephasing of ^{T}Mu by precession; especially if P_R is to equal the primary muonium yield (h_M). (ii) Perhaps such reactions may be assisted by tunnelling in the case of Mu, but this would lead to somewhat less selectivity than in the case of H-atom reactions. (iii) Addition to the π-bond may occur during the epithermal stage by a hot atom reaction of Mu*. Some selectivity is possible in the hot-atom step, but a number of primary radicals would probably be formed - such as vinyl cyclohexadienyl radicals in the case of styrene. If these simply convert to the most stable radical by intramolecular rearrangements with high probability then $P_R \sim h_M$. If the less-stable radicals decompose or react, rather than rearrange, then P_R will be smaller than h_M. Finally, (iv) radical formation could result from a μ^+-adduct, without involving Mu, by yielding a radical cation capable of being neutralized by quasi-free electrons or radical anions at the end of the muon track. Radical cations undergo rapid internal rearrangements, so complete selectivity is also possible via ionic reactions.

Perhaps some of these alternatives can be ruled out on the basis of the measured radical yields. Fourier spectra provide the power level at each radical frequency, with power being proportional to amplitude squared. In evaluating P_R one has to sum the amplitude over all radical frequencies. In the high-field limit this means just two frequencies per radical; but with more than one radical present then P_R, as used in chapter 6, should be equated with the total radical yield, $\Sigma P_i(R)$. There are two major sources of inaccuracy in determining P_R, plus various possible reactions - such as intratrack processes - which can lead to lost polarization prior to the μSR measurement at $\sim 10^{-6}$ s. The first source of inaccuracy arises from the fact that standard μSR technology cannot fully respond to the frequencies of the radical precessions. Roduner *et al.* (1982*b*) have recently measured the response-time of their equipment, as shown in Figure 9.8. This enabled them to make corrections to the amplitudes through Eq. [9.5],

$$A^{obs} = A^{true} \exp[-(\pi \nu \Delta t)^2 / 4 \ln 2] \qquad [9.5]$$

following Holzschuh, Kundig & Patterson (1981), where Δt is the full width at half maximum, with the time resolution assumed to be Gaussian, and ν is the observed frequency. It means, for their equipment, that the observed amplitudes are down by a factor of 1.4 at $\sim$200 MHz and 1.8 at $\sim$300 MHz. These are obligatory corrections to make in yield measurements and obviously affect the benzene more than the styrene in Figure 9.7, for instance. The second inaccuracy stems from the old problem of a dephasing of the muon's spin by precession in a prior muon state, due to the differences in frequencies of the initial and final states being comparable to the reaction rate. This problem has been tackled theoretically by Percival & Hochmann (1979), who showed how its magnitude depends on the rate constant and magnetic field. Complete and equal amplitudes for the two radical frequencies occur only when $\lambda \geqslant 10^{11}\ \mathrm{s}^{-1}$ at normally used fields. For a liquid of molar concentration ~10, this means the value of k_M must exceed $10^{10}\ \mathrm{M}^{-1}\ \mathrm{s}^{-1}$ or corrections are necessary. Whereas many π-bond addition reactions of Mu proceed on the first encounter, the competition invoked in (i) above, and the intermolecular selectivity in the benzene-styrene mixtures, both imply much slower reactions. However, without knowledge of the actual rate this latter correction cannot be applied. It is fortunate that both these inaccuracies reduce the higher frequency more than the lower one. Thus if the observed difference corresponds to that predicted by Eq. [9.5] then it suggests dephasing can be neglected.

As a result of these problems, P_R values are generally less accurate than P_D and P_M. Nevertheless, Roduner *et al.* (1982*b*) have obtained some interesting results, a selection of which is provided in Table 9.2. There is a missing fraction

Figure 9.8. Calibration curve for corrections to be applied to the amplitudes of Fourier transform spectra due to the finite time resolution of the μSR equipment. Broken lines are theoretical curves based on Eq. [9.5] for Δt values of 1.3 ns (upper) and 1.5 ns (lower). (Figure taken from Roduner *et al.*, 1982*b*.)

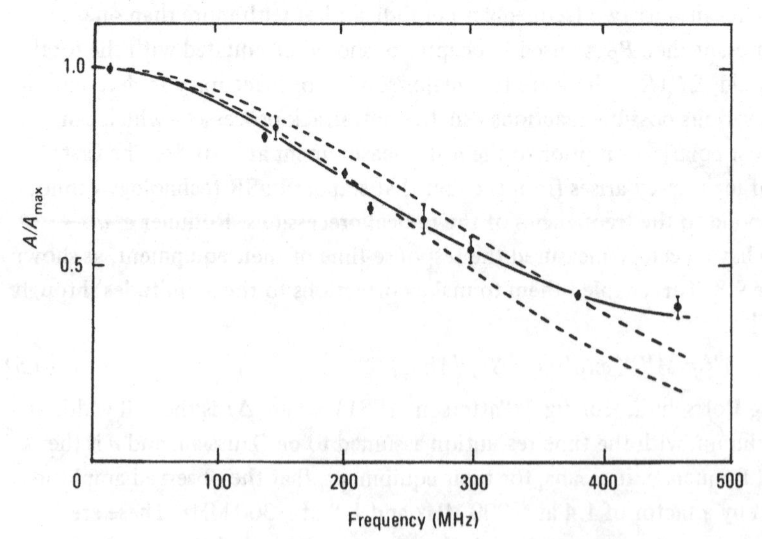

(P_L) of 0.2 for benzene where only one radical is possible. (Virtually the same missing fraction occurs in polar liquids and saturated hydrocarbons where Mu is observed rather than radicals (Ito, Ng, Jean & Walker, 1980).) Even in substituted benzene when two or more radicals are found and their individual yields are quite different, P_L is still quite small - often in the range 0.2 to 0.35. This suggests that the radical-formation mechanism is extremely rapid. At the same time selectivity is efficient.

9.6 Reactions of Mu-radicals

Increasing linewidths in frequency-space corresponds to decreasing lifetimes in time-space. Therefore, with sufficient resolution in the precession frequency measurements, one can obtain the rate of decay of the radical directly from a plot of the change in linewidth in a Fourier spectrum as a function of substrate concentration. This is easier than trying to evaluate the change with time of the amplitude at one particular radical frequency. A rate constant of $2.6 \times 10^8 \mathrm{M}^{-1} \mathrm{s}^{-1}$ has already been determined for the reaction of the well-known radical scavenger *p*-benzoquinone with the Mu-cyclohexadienyl radical formed in benzene (Roduner, 1981*b*). There is also the possibility of measuring some of the faster propagation rates in vinyl monomers by observing the change in linewidth of the Mu-radical as the monomer is diluted in an inert solvent.

Muonium is merely the means of monitoring the radical. It is attached at least one C atom away from the radical centre, so it should not influence the chemical interaction perceptibly nor create an isotope effect in these radical reaction rate measurements.

Table 9.2. *Radical yields as a fraction of the total muon polarization (P_R for each radical), together with P_D values, for benzene, toluene and fluorobenzene at ~295 K*[a]

Parent molecule	Radical position[b]	A (MHz)	P_R	P_D	P_L[c]
benzene	–	514.6	0.65	0.15	0.20
toluene	2	489.6	0.239	0.25	0.25
	3	509.3	0.086		
	4	496.4	0.174		
fluorobenzene	2	485.7	0.18	0.19	0.35
	3	511.8	0.06		
	4	511.8	0.19		

(*a*) Data selected from that of Roduner *et al.*, 1982*b*.
(*b*) Taking CH_3 or F at position 1 on the ring.
(*c*) $P_L = 1 - P_D - \Sigma P_R$ since $P_M = 0$ in these aromatic compounds.

9.7 Radical studies in low magnetic fields

Figures 9.1 and 9.2 show how widely the muon polarization is dispersed over many frequencies due to the nuclear Zeeman effect, even from just one other nuclear spin, when low transverse magnetic fields are used. But such distributions and splittings may well be utilized, when needed, to characterize radical structures or to identify coupling constants.

Splitting of radical frequencies for identification purposes can be demonstrated even by using moderately high magnetic fields. Thus, the benzene lines separated by 514 MHz are each split by 1.5 (±0.2) MHz as seen in Figure 9.9, at 1000 G due to the α H (Roduner & Fischer, 1981). These authors have shown that at low fields the treatment becomes very complicated with two or more protons, but that at zero field the situation is easily analysed – particularly when there are present just a few magnetically equivalent nuclei other than the muon. At zero fields many of the lines become degenerate, as they do in free Mu. As Percival (1981) has pointed out, a free radical containing the muon but no other nuclear moment ($Mu\dot{O}$, perhaps) would be indistinguishable from free Mu at fields too low to split the 1.39 MHz G^{-1} precession.

Figure 9.9. μSR precession frequencies obtained with benzene in transverse fields of 4 kG (upper spectrum) and 1 kG (lower spectrum). The 1.5-MHz splitting in the latter case is readily discernible. (Figure taken from Roduner & Fischer, 1981.)

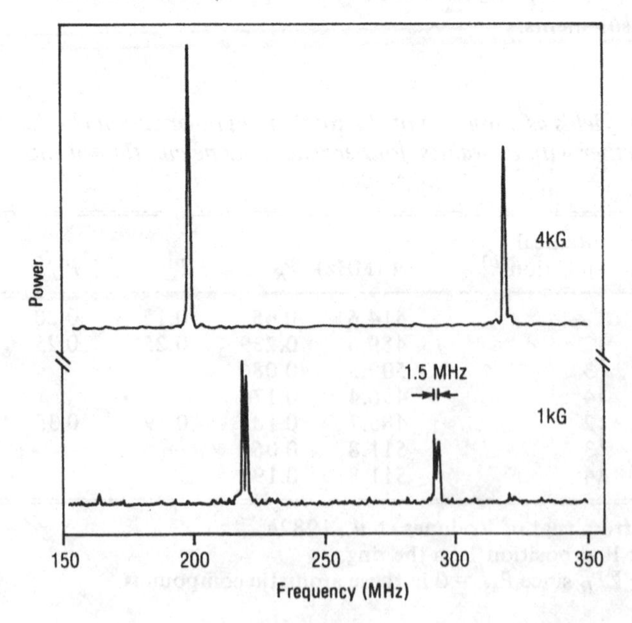

9.8 Mu-radicals in single crystals

Finally, μSR can be used to study the three-dimensional anisotropy of the muon's hyperfine interaction tensor in different orientations of single crystals. Roduner (1981*a*) has demonstrated this using durene (1,2,4,5-tetramethylbenzene), a molecule with a centre of symmetry which forms monoclinic crystals. He measured the intensities of the two pairs of radical frequencies in the high-field limit for each 30° of rotation about the three major axes of the crystal.

This type of analysis shows the relative simplicity of μSR compared with ESR, as the information obtained from this highly complex radical concerns the interactions of only one of its nuclei - the spin-polarized muon.

References for chapter 9

Brewer, J. H., Crowe, K. M., Gygax, F. N., Johnson, R. F., Fleming, D. G. & Schenck, A. (1974). Muonium chemistry in liquids: evidence for transient radicals. *Physical Reviews*, **A9**, 495–507.

Brewer, J. H., Fleming, D. G. & Percival, P. W. (1982). Fourier transform μSR. In *Fourier, Hadamard, and Hilbert Transforms in Chemistry*, ed. A. G. Marshall, pp. 345–85. New York: Plenum Publishing Corporation.

Brodskii, A. M. (1963). Conditions of formation of μ^+ mesic molecules. *Soviet Physics, JETP*, **17**, 1085–8.

Bucci, C., Guidi, G., De'munari, G. M., Manfredi, M., Podini, P., Tedeschi, R., Crippa, P. R. & Vecli, A. (1978). Direct evidence for muonium radicals in water solutions. *Chemical Physics Letters*, **57**, 41–4.

Fessenden, R. W. (1964). ESR studies of internal rotation in radicals. *Journal de Chimie physique*, **61**, 1570–5.

Fischer, H. & Hellwege, K.-H. (1977–79). Editors: *Magnetic Properties of Free Radicals*. Landolt-Bornstein, New Series, Group II, vol. **9**. Berlin: Springer-Verlag.

Heller, C. & McConell, H. M. (1960). Radiation damage in organic crystals. *Journal of Chemical Physics*, **32**, 1535–9.

Holzschuh, E., Kundig, W. & Patterson, B. D. (1981). Direct measurement of the hyperfine frequency of muonium in silicon. *Hyperfine Interactions*, **8**, 819–21.

Ito, Y., Ng, B. W., Jean, Y. C. & Walker, D. C. (1980). Muonium atoms observed in liquid hydrocarbons. *Canadian Journal of Chemistry*, **58**, 2395–401.

Ng, B. W. *et al.* (1982). To be published.

Percival, P. W. & Fischer, H. (1976). Theory and analysis of μ^+ spin polarization in chemical systems. *Chemical Physics*, **16**, 89–99.

Percival, P. W. & Hochmann, J. (1979). Formation of muonic radicals. *Hyperfine Interactions*, **6**, 421–4.

Percival, P. W. (1981). Use of spin polarized muons to probe hydrogen atom reactions. *Proceedings 15th International Free Radical Symposium*, Nova Scotia.

Roduner, E., Percival, P. W., Fleming, D. G., Hochmann, J. & Fischer, H. (1978). Muonium-substituted transient radicals observed by muon spin rotation. *Chemical Physics Letters*, **57**, 37–40.

Roduner, E. & Fischer, H. (1979). The evolution of muon spin polarization in muonic radicals and related species. *Chemical Physics Letters*, **65**, 582–6.

Roduner, E. (1979). *On the Liquid Phase Chemistry of the Light Hydrogen Isotope Muonium*, Ph.D. Thesis, University of Zurich, pp. 1–97.

Roduner, E. (1981*a*). Observation of muonium substituted free radicals in a durene single crystal. *Chemical Physics Letters*, **81**, 191–4.

Roduner, E. (1981*b*). Muonium-substituted free radicals. *Hyperfine Interactions*, **8**, 561–70.

Roduner, E. & Fischer, H. (1981). Muonium substituted organic free radicals in liquids. Theory and analysis of μSR spectra. *Chemical Physics*, **54**, 261–76.

Roduner, E., Strub, W., Burkhard, P., Hochmann, J., Percival, P. W., Fischer, H., Ramos, M. & Webster, B. C. (1982*a*). Muonium substituted organic free radicals in liquids. Muon-electron hyperfine coupling constants of alkyl and allyl radicals. *Chemical Physics*, **67**, 275–85.

Roduner, E., Brinkman, G. A. & Louwrier, W. F. (1982*b*). Muonium substituted organic free radicals in liquids. Muon-electron hyperfine coupling constants and the selectivity of formation of methyl and fluorine substituted cyclohexadienyl type radicals. *Chemical Physics* (submitted).

Stadlbauer, J. M., Ng, B. W., Walker, D. C., Jean, Y. C. & Ito, Y. (1981). Muonium addition to vinyl monomers. *Canadian Journal of Chemistry*, **59**, 3261–6.

Added in proof: Two further references of interest which have just appeared are:

Hill, A., Allen, G., Stirling, G. & Symons, M. C. R. (1982). Muon spin rotation studies of organic liquids. *Journal of the Chemical Society, Faraday Transactions I*, **78**, 2959–73.

Cox, S. F. J., Hill, A. & De Renzi, R. (1982). Chemical and magnetic properties of muonic free radicals in solutions of styrene in benzene. *Journal of the Chemical Society, Faraday Transactions I*, **78**, 2975–95.

10

MUONIC ATOMS - THE CHEMISTRY OF μ^-

10.1 Preamble

To this point, only the chemistry of μ^+ (which is the antiparticle) has been discussed and it was seen to behave as if it were a very light proton. Now it is the turn for the muon itself (μ^-) which emulates the properties of a super-heavy electron. It takes on an 'orbital' rather than a 'nuclear' role. During its short lifetime μ^- gets caught in the coulomb field of a nucleus, replacing an electron to form a muonic atom - the simplest of which is muonic hydrogen ($p^+\mu^-$). Being a lepton, the muon experiences no strong nuclear interaction, only the weak interaction governing its natural decay and the electromagnetic interaction controlling atomic properties.

A glance at the Bohr picture of the hydrogen atom (and other one-electron systems) as in Eq. [10.1],

$$r \propto n^2/Zm^* \quad \text{and} \quad E_n \propto Z^2m^*/n^2 \qquad [10.1]$$

reminds one that the mean radius of the orbit of principal quantum number n is inversely proportional to the reduced mass (m^*), and its energy is directly proportional to that mass. With a muon (207 times the mass of an electron) replacing the electron in hydrogen, m^* increases by a factor of 186. So the $n = 1$ orbit in muonic hydrogen is only 2.8×10^{-13} m away from the nucleus compared to 5.3×10^{-11} m for ordinary H. Furthermore, the 'ionization energy' is 186 times greater. This fantastic difference in size and binding energy means that muonic and ordinary hydrogen should not be considered 'isotopes' in any sense of the word: there is virtually no chemical resemblance between them. Instead, muonic hydrogen is like a short-lived free neutron.

When μ^- replaces an electron in an atom of atomic number Z, then the muonic ground state (1s) orbital is even closer to the nucleus than it is in ($p^+\mu^-$), as indicated by the Z in Eq. [10.1]. One can still apply the Bohr picture here because this muonic 1s orbital is almost completely inside the remaining electron orbitals of the atom: they are hardly influencing it, and it alters their energy merely by reducing the effective nuclear charge to $(Z-1)$. For atom Z, the 1s muonic Bohr radius (r_Z) is thus related to that of H (r_1) by Eq. [10.2],

$$r_Z = r_1/Z(m_\mu^*/m_e^*) \quad [10.2]$$

where (m_μ^*/m_e^*) is the reduced mass ratio for μ^- compared to e^-. With $Z=79$ (Au), for instance, the muon in muonic gold is most likely to be found only 3.2×10^{-15} m from the nuclear centre, which is comparable to the size of the nuclear field. Certainly the muon and nucleon wave functions overlap so strongly that the muon spends a significant fraction of its time in the space occupied by the nucleus. So much so that the muon's energy is strongly influenced by the actual distribution of charge within the nucleus, by nuclear quadrupole moments, polarization susceptibilities, and by other nuclear properties: which makes a muonic atom a particularly valuable probe of nuclear structure.

However, the initial atomic capture probability is a strong function of the actual electronic structure (chemical composition) of the material being studied. So muonic atoms have potential use in chemistry too - as both qualitative and quantitative analytical tools of molecular structure. As will be seen, they provide an identification signature in the form of a characteristic X-ray spectrum or lifetime, and a quantitative measure in the form of intensity.

Muonic atoms are sometimes categorized together with mesonic (or mesic) atoms comprising a negative meson, such as π^- or K^-, in an orbital role. But there are two basic differences. First, the mesons undergo strong (nuclear) interactions, therefore they get assimilated by the nucleus from high atomic orbitals and never reach the 1 s ground state: so their lifetimes are greatly reduced and they do not probe the nucleus as non-interacting particles. Second, they are spinless particles, so the total angular momentum is given only by $j=1$. By contrast, the muon is a spin 1/2 particle, so the individual levels of muonic atoms are doublets and the same n, l quantum numbers and selection rules apply as in electronic atoms.

10.2 Atomic capture and X-ray production

There are two distinct stages of capture to consider: first the initial trapping into an orbital of the atom or molecule, and its subsequent relaxation; and second, the possibility of eventual nuclear capture and disappearnace. The probability of the first - the atomic capture step - gives rise to the yield of muonic atoms. Its relaxation generates the X-ray emission and it is the nuclear capture step which regulates its lifetime. In light atoms nuclear capture barely competes with the 2 μs spontaneous decay of the muon, but in heavy atoms it dominates (see later).

Muon capture by atoms was first demonstrated in some classical experiments by Conversi, Pancini & Piccioni (1947), almost concurrent with the formulation of the Fermi–Teller Z-law (1947). According to this law, the muon capture probability would show a simple dependence on the number of electrons in an atom and therefore would be proportional to Z. However, this law was soon found not to hold at all well when considered in detail. It overestimates the capture rate in heavy atoms (Goulard & Primakoff, 1974); does not account for an isotope effect and the fact that odd-Z atoms have higher capture rates than even-Z (Eckhause, Siegel, Welsh & Filippas, 1966; Suzuki, 1980); nor for a periodicity following the Periodic Table with minima at the alkali metals (Sens, Swanson, Telegdi & Yovanovitch, 1957, 1958; Eckhause *et al.*, 1962; Baijal, Diaz, Kaplan & Pyle, 1963); nor can it be used to account for a strong dependence on molecular composition (Zinov, Konin & Mukhin, 1966; Schneuwly *et al.*, 1978*a*; Schneuwly, Pokrowsky & Ponomarev, 1978*b*); nor, finally, for a subtle variation with molecular structure (Ponomarev, 1973; Knight *et al.*, 1980). It was therefore subjected to considerable modification, particularly by Vasilyev *et al.* (1976), Schneuwly *et al.* (1978*a*,*b*) and Daniel (1979). Thus electronic structure was incorporated by distinguishing core from valence electrons and by including an inverse dependence on atomic radii. One such relationship is given in Eq. [10.3] (Daniel, 1979),

$$w(Z_1/Z_2) = (Z_1/Z_2)^{1/3}(\ln 0.57Z_1/\ln 0.57Z_2)(R_2/R_1) \qquad [10.3]$$

where $w(Z_1/Z_2)$ is the ratio of capture probabilities for atoms of atomic number Z_1 and Z_2, with R_1 and R_2 their atomic radii.

Some measure of the current status can be gleaned from Figure 10.1 which shows the capture probability for metal atoms relative to oxygen in oxides. This Figure includes experimental data from three broad studies and also gives the lines corresponding to three postulated relationships (Suzuki, 1980). Though hardly at the point of satisfactory resolution, there is evidently a periodicity corresponding to chemical influences. It is not clear whether the chemical effects can be described better as perturbations to the basic atomic capture process – utilizing valence electron structure and atomic radii as in [10.3] (Vogel, Haff, Akylas & Winther, 1974; Daniel, 1979) – or whether the chemical bond should be taken into account directly, as if the initial trapping is in a molecular orbital (Gershtein, Petrukhin, Ponomarev & Prokoshkin, 1969; Schneuwly *et al.*, 1978*b*).

In any event, the dependence on electronic structure is established and this is consistent with capture occurring at relatively low μ^- energies. Capture is now believed to occur primarily while the muon has only 10–20 eV or less of kinetic energy (Korenman & Rogovaya, 1980) where it can be more selective and discriminating than at the hundreds of eV anticipated at the time the Fermi-Teller law was formulated.

Most muonic atom studies utilize high energy ('backward') μ^- beams coming from π^- decay (perform CP transformation on Figure 1.1). The muon loses most of its energy by collisions with the atoms or molecules of the medium by causing ionization and excitation. Unlike μ^+ it does not undergo charge-exchange as it reaches the velocity of the atomic electrons, but before dropping to thermal energy it is invariably trapped by the coulomb field of an atom. Capture initially occurs in a high Rydberg state of the atom. According to an adiabatic capture mechanism (Wightman, 1950) the approaching μ^- partially screens the nuclear charge so that an electron is lost and replaced by the muon. For H-atoms this takes place with maximum probability in an orbit for which $n \sim 0.8\,(m^*)^{1/2}$, so that the capture cross-section peaks at $n \sim 14$ for μ^-. [For π^- and K$^-$ the

Figure 10.1. Muon capture ratios $W(Z/O)$ in various oxides ZO, where Z is the atomic number. Experimental data points are from: ⊙, Suzuki (1980); ⍙, Daniel, 1979; ×, Zinov *et al.* (1966). Experimental errors for the latter are not given in the Figure to diminish cluttering; but they are comparable to the other two sets of data points. The lines represent various theoretical attempts to account for the results over certain regions of Z. The solid line is the basic modified Z-law of Vasilyev; the dashed line corresponds to Daniel's treatment, following Eq. [10.3]; and the dotted lines correspond to the calculations of Schneuwly *et al.* (1978*b*).

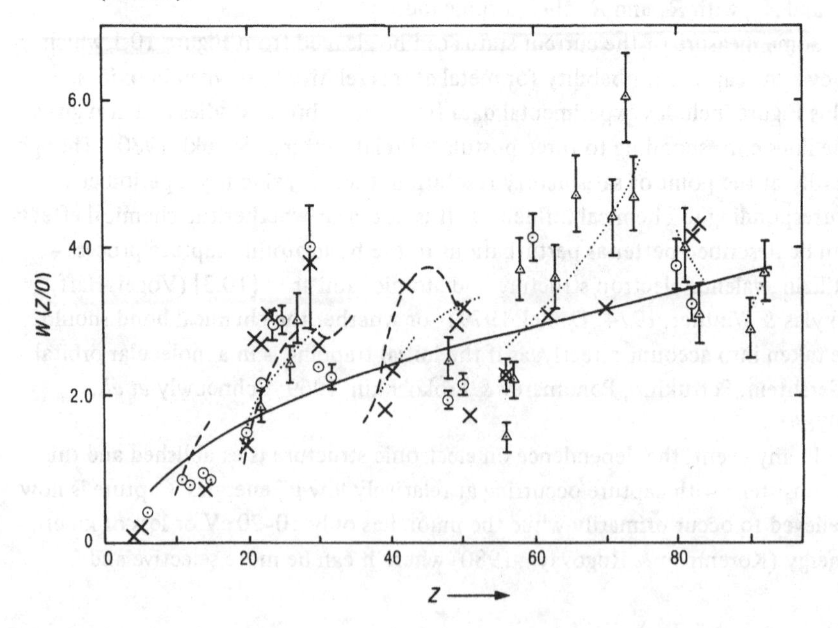

corresponding mesic H-atoms have n values of ~15 and ~26, respectively.] Transitions to lower muon energy states then proceed either by X-ray emission or by replacing electrons (Auger transitions). Selection rules of $\Delta l = \pm 1$ and $\Delta m = 0, \pm 1$ apply to both processes; and the probability depends on $\sim(\Delta E)^3$ for the radiative route and on $\sim(\Delta E)^{-1/3}$ for the Auger emission. This gives the latter route its greatest importance between higher excited states of light atoms. The overall cascade results in a simple X-ray emission spectrum, with the last transition (2p → 1s) corresponding to the lowest energy of the Lyman (K shell) lines being the most prominent and the most widely used. A muonic atom is readily identified by its X-ray spectrum, with Figure 10.2 showing the kind of data obtained. One determines the number of atoms from the sum of the intensities of the K-series of lines after appropriate normalizations and corrections for self-absorption. An Auger electron spectrum could also be used for the same purpose.

When molecules rather than atoms are studied, the initial step has been described as capture into a molecular orbital to give highly excited muonic molecules (Ponomarev, 1973; Gershtein & Ponomarev, 1975). Here, the mean muon-nuclear distance greatly exceeds the muonic atom radii discussed earlier, but it should be responsive to the molecule's electronic structure. Localization

Figure 10.2. Lyman series (K shell) X-ray emission from the cascade de-excitation transitions of μ^- stopped in Ti. Note that the 2p → 1s transition is an order of magnitude more intense than shown. The calculated intensities were derived from cascade calculations and corrected for target self-absorption and detector sensitivity. (Figure taken from Kessler *et al.*, 1967.)

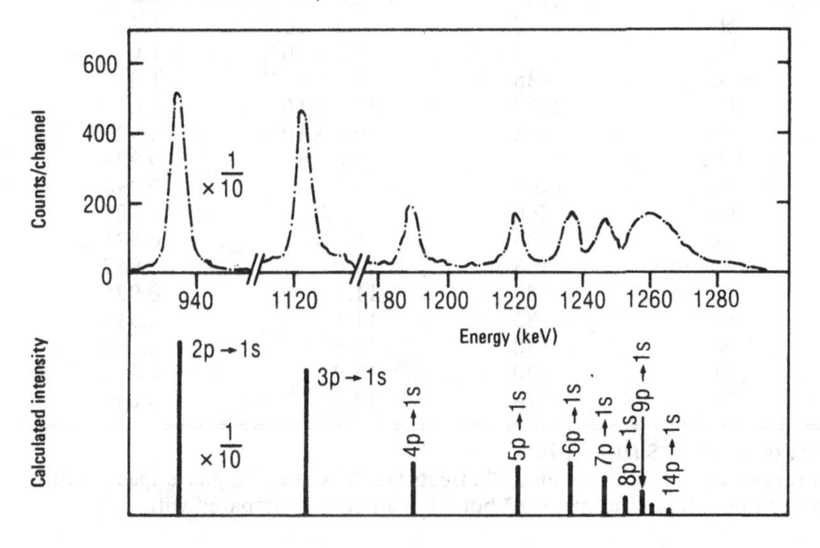

into atomic orbitals close to the various nuclei rapidly follows, with a distribution of probabilities embodying a message about the molecular structure. So the spectrum of the cascade X-ray intensities is affected by chemical bonds (Zinov *et al.*, 1966; Kessler *et al.*, 1967; Cox *et al.*, 1979).

10.3 Muonic atom lifetimes and nuclear capture

The theorem which collectively conserves CPT (charge-parity-time inversions) requires the natural lifetime of μ^- in Eq. [10.4]

$$\mu^- \rightarrow e^- + \nu_\mu + \bar{\nu}_e \qquad [10.4]$$

to be the same as that of μ^+ (2.197 μs) – in free space. However, one is seldom concerned with muons *in vacuo* since they invariably come to rest in muonic atoms. When bound within such an atom, the muon decay rate is somewhat faster than the free value due to the reduced phase space (proportional to the fifth power of the fractional difference between the muon's mass and binding energy) and due to Doppler and coulomb effects – all collectively described by the Huff factor (Suzuki, 1980: see Table 10.1). But in heavy atoms in particular there is a much more important factor, namely nuclear capture. It was noted initially by Conversi *et al.* (1947) that the decay electrons (Eq. [10.4]) were

Table 10.1. *Selection of muonic atom lifetimes, together with their nuclear capture rates and Huff correction factors*[(a)]

Z	Element (isotope)	Mean lifetime (ns)	Nuclear capture rate ($\times 10^6\,s^{-1}$)	Huff factor[(b)]
Free space		2197	–	–
1	H	2195	5×10^{-6}	1.0
3	^{6}Li	2177	4.2×10^{-3}	1.0
	^{7}Li	2188	1.8×10^{-3}	1.0
6	C	2027	13.8×10^{-2}	1.0
8	^{16}O	1795	10.4×10^{-2}	0.998
	^{18}O	1844	8.8×10^{-2}	0.998
11	Na	1204	0.377	0.996
17	Cl	560	1.33	0.989
26	Fe	206	4.41	0.975
35	Br	133	7.07	0.952
47	Ag	87	11.1	0.925
53	I	83	11.6	0.910
74	W	78	12.4	0.860
81	Tl	70	13.9	0.846
83	Bi	74	13.1	0.840

(a) Data taken from Suzuki, 1980.
(b) Huff correction factors include all effects (such as reduced phase space) which reduce the natural lifetime of μ^- bound in an atom, compared with free space.

generated on a shorter timescale when μ^- was stopped in iron compared with carbon. In fact in Fe no electrons were detected after ~1 μs. This led to the concept of nuclear capture, orders of magnitude slower than the true strong interaction, corresponding to Eq. [10.5] for free protons,

$$\mu^- + p^+ \rightarrow n + \nu_\mu \quad [10.5]$$

and to Eq. [10.6] for protons bound in complex nuclei.

$$\mu^- + (A, Z) \rightarrow (A, Z-1) + \nu_\mu \quad [10.6]$$

Products of reaction [10.6] are neutron-rich excited nuclear states which generally undergo relaxation by neutron, but not electron, emission. The capture cross-section increases with Z, as the 1s orbital overlaps the nucleus more and more (see Table 10.1).

The overall disappearance rate of μ^- is governed by the sum of these two terms: the natural decay rate, modified by the phase space, plus the nuclear capture rate. Heavy atoms have shorter muonic lifetimes than lighter ones, as can be seen by the selection provided in Table 10.1. They range from virtually the natural τ_μ in H (2.195 μs) to 0.070 μs in Tl.

Figure 10.3. Lifetime spectrum of μ^- in Cr_2O_3 powder, plotted as the number of decay-electron counts as a function of time. The data are deconvoluted through Eq. [10.7] into four components, *Cr*, *O*, *C* and background (*Bg*). (Figure taken from Suzuki *et al.* (1980).)

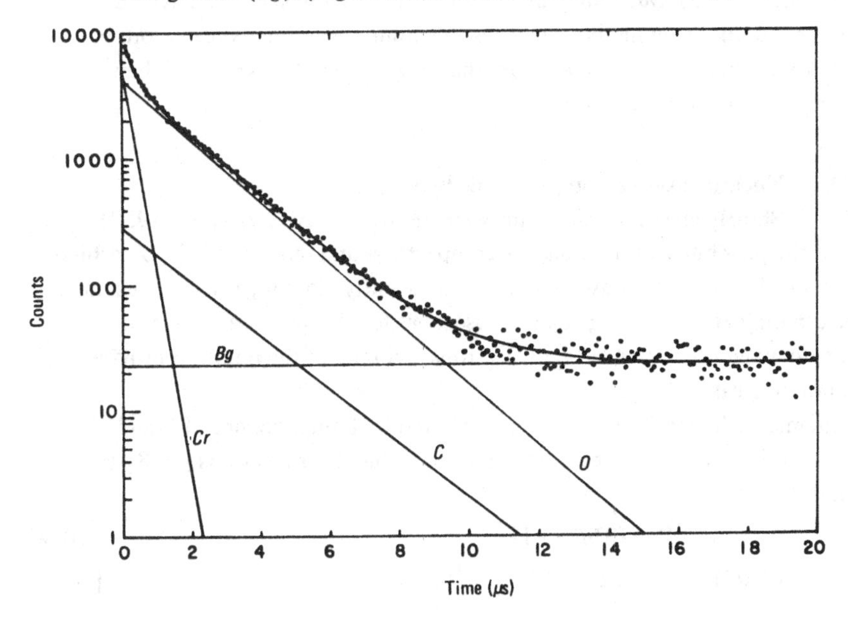

Some of the first muonic atom studies utilized lifetime measurements to identify and enumerate the atoms formed (Sens *et al*., 1958; Lathrop *et al*., 1960). The lifetimes were evaluated from decay electron histograms – decay rates as a function of time. This method requires that the lifetimes of the various muonic atoms be sufficiently well separated that an accurate deconvolution can be made. Figure 10.3 gives a clear-cut recent example for the case of Cr_2O_3 (Suzuki *et al*., 1980). This histogram was fitted to Eq. [10.7]

$$A(t) = \Sigma A_i \exp(-t/\tau_i) + Bg \qquad [10.7]$$

where Bg is a time-independent background count rate. A and τ are the amplitude and mean lifetime of the various elements present, which in this case were Cr, O and C – the latter from the hydrocarbon polymer scintillation counters and from the walls of the cell containing the Cr_2O_3 powdered sample. These amplitudes have to be corrected for the fraction of muonic atoms decaying via Eq. [10.4] rather than [10.6], and for various electron-loss processes. The Cr lifetime ($\tau_{Cr} = 0.26\,\mu s$) is so much shorter than that of O ($\tau_O = 1.80\,\mu s$) that its intensity looks low; but its correction factor makes up for this because its decay is largely through nuclear capture (~88%). There is good agreement in general between relative capture rate data obtained by the lifetime and X-ray methods; but absolute cross-sections cannot yet be determined with confidence.

It has been suggested that the different stopping ranges of muons of different energies might be employed to scan the chemical constitution of a large object as a function of distance – for example, to map the chemical structure of man (Ponomarev, 1973). But it may be some time before the analytical 'resolution' of these muonic atom methods reaches the point where the oxygen atom in 1-propanol can be distinguished from that in 2-propanol by virtue of their different capture cross-sections.

10.4 Nuclear fusion through muonic hydrogen

Shortly after muonic atoms were discovered, Alvarez *et al*. (1957) tested the possibility of utilizing the compactness and tenacity of the μ^- orbital in muonic hydrogen to draw two nuclei sufficiently close together for fusion. One can only envisage using H and its isotopes for this purpose because one wants to avoid the presence of all electrons and rely on the muon to provide the internuclear binding.

Muonic hydrogen ($p^+\mu^-$) may be generated by bombardment of liquid H_2 by energetic muons and used as the starting material for reactions [10.8] and [10.9],

$$(p^+\mu^-) + {}^2H \rightarrow (d^+\mu^-) + H \qquad [10.8]$$

$$(d^+\mu^-) + H \rightarrow (d^+\mu^-p^+) + e^- \qquad [10.9]$$

where d is the deuteron. The internuclear separation in $(d^+\mu^-p^+)$ is estimated to be only $\sim 2\times 10^{-13}$ m, and at that distance tunnelling leads to Reaction [10.10]

$$p^+ + d^+ \rightarrow {}^3He^{2+} + \gamma + 5.5\,\text{MeV} \qquad [10.10]$$

with a significant probability. These basic reactions were proved to occur; but the muon lifetime is $\sim 10^2$-fold too short for this to be of any technological use.

It has also been noted that a typical meson factory consumes about a megawatt of power for every watt of muons produced. Therefore recycling of the muons is essential: perhaps in some form of chain reaction in which μ^- merely acts as the binding catalyst. There is also a problem here with 'sticking' of μ^- to the product He at about the 1% level, which seems to restrict chain lengths to ~100. However, the possibility of harnessing muonic ^1H, ^2H and ^3H to give a positive energy gain continues to be studied experimentally and theoretically and should not yet be ruled out (Petrov, 1980).

10.5 μ^-SR

In principle, μSR can be applied equally well to studies of μ^- as to μ^+, since the particle is initially polarized and electron emission is directed preferentially against the muon's spin direction; but in practice the signals are much weaker, for two reasons. First, the nuclear capture process both reduces the mean lifetime and allows electron-emission from only a fraction of the muon ensemble. Second, the residual polarization of muons in muonic atoms is greatly reduced. This latter effect arises from the succession of capture events. Atomic capture itself causes the spin polarization to be reduced to about one-third; then the cascade of transitions to the 1s state causes a further loss of about one-half (producing partially circularly polarized X-radiation, incidently). There is a further depolarization due to hyperfine coupling if the nucleus has a non-zero spin. In muonic ^{19}F, for instance, only ~4% of the initial polarization is retained, in muonic ^{12}C about 18% (Suzuki, 1980).

Despite this greatly reduced sensitivity, μ^-SR remains a vital source of information on muonic atoms. The extent and nature of the depolarization processes during capture and cascade, as mentioned above, were studied mainly by measurement of the residual polarization by μ^-SR in a variety of materials (Ignatenko, 1961; Evseer, 1975). Furthermore, the spin precession and relaxation of the μ^- polarization of a bound muonic atom in its ground state can be used to explore the electronic structure and spin density just outside the nucleus for a variety of interests. It is also an extended nuclear probe, behaving as a dilute quasi-nucleus of atomic number $(Z-1)$ whose effect reaches well beyond the nucleus. Thus it has found use mainly in solid state physics: to monitor local magnetic fields, to explore the hyperfine anomaly, and to measure paramagnetic shifts (Yamazaki *et al.*, 1974; Nagamine, 1979; Takagi *et al.*, 1981).

10.6 Pionic hydrogen

This is the simplest and most studied of the mesonic atoms, which offers some useful comparisons with muonic atoms. Pionic hydrogen ($p^+\pi^-$) forms very readily when negative pions are stopped in a material containing hydrogen, through an Auger capture mechanism (Leon & Bethe, 1962; Ponomarev, 1973). The small neutral highly excited $p^+\pi^-$ atom then dissociates itself from the molecule, loses its excess energy by collision with the molecules of the medium, and interacts with them (particularly with their H atoms); or its pion is transferred to a heavier atom leading to nuclear capture and disintegration. Its lifetime is less than one picosecond. Pionic hydrogen's existence is detected by the observation of 2γ-rays of ~70 MeV from the charge-exchange reaction [10.11],

$$p^+\pi^- \rightarrow n\pi^\circ \rightarrow 2\gamma + n \qquad [10.11]$$

which occurs by interaction with a further proton.

The probability that a pion which stops in a sample containing H will be absorbed by a proton is closely related to the properties of the chemical bond and especially to its covalency. There is a high degree of selectivity. This has been amply demonstrated with oxyacids, bases, organic compounds, metal hydrides, crystalline hydrates, and water at various temperatures and states of hydrogen bonding (Horvath, 1981).

Pionic hydrogen is already developed as a useful analytical tool because of its efficient generation as a neutron-like dynamic reagent and its coincident 2γ signature. When negative pions are captured by other atoms, the strong interaction is so pervasive that nuclear capture occurs before the cascade of radiation or Auger transitions is complete. π^- is generally lost before dropping to $n = 4$. The lifetime of pionic atoms is thus on the 10^{-12}-s scale despite the free π^- living for 26 ns.

Negative pion beams are also being utilized in some of the applied programs at meson facilities for cancer therapy (Skarsgard, Henkelman, Lam & Poon, 1979). The principal advantage of this type of radiation for the treatment of deep-seated tumours stems from the high fraction of its total energy which is deposited in the last 10–20% of its track. This 'star effect' arises from nuclear capture and disintegration as the pion disappears following pionic-atom formation. In addition to its superior geometric deposition, this star radiation is of high linear-energy-transfer: so there is a greatly reduced oxygen-enhancement-effect, with the result that tumour cells which have a deficient oxygen supply are not less sensitive to radiation damage than healthy ones.

References for chapter 10

Alvarez, L. W., Braidner, H., Crawford, F. S., Crawford, J. A., Falk-Vairant, P., Good, N. L., Gow, J. D., Rosenfeld, A. H., Solmitz, F. T., Stevenson, M. L., Ticho, H. K. & Tripp, R. D. (1957). Catalysis of nuclear reactions by μmesons. *Physical Review*, **105**, 1127–8.

Baijal, B. J. S., Diaz, J. A., Kaplan, S. N. & Pyle, R. V. (1963). Atomic capture of μ^- mesons in chemical compounds. *Nuovo Cimento*, **30**, 711–26.

Conversi, M., Pancini, E. & Piccioni, O. (1947). On the distintegration of negative mesons. *Physical Review*, **71**, 209–10.

Cox, C. R., Dodson, G. W., Eckhause, M., Hart, R. D., Kane, J. R., Rushton, A. M., Siegel, R. T., Welsh, R. E., Carter, A. L., Dixit, M. S., Hincks, E. P., Hargrove, C. K. & Mes, H. (1979). Muonic X-ray intensities in low-Z elements and their hydrides. *Canadian Journal of Physics*, **57**, 1746–8.

Daniel, H. (1979). Coulomb capture of muons and atomic radius. *Zeitschrift für Physik A*, **290**, 29–31.

Eckhause, M., Filippas, T. A., Sutton, R. B., Welsh, R. E. & Romanowski, T. A. (1962). Experimental test of the Fermi–Teller Z-law. *Nuovo Cimento*, **24**, 666–71.

Eckhause, M., Siegel, R. T., Welsh, R. E. & Filippas, T. A. (1966). Muon capture rates in complex nuclei. *Nuclear Physics*, **81**, 575–84.

Evseer, V. S. (1975). Depolarization of negative muons and interaction of mesonic atoms with the medium. In *Muon Physics*, vol. **3**, ed. V. W. Hughes & C. S. Wu, pp. 235–98. New York: Academic Press.

Fermi, E. & Teller, E. (1947). The capture of negative mesotrons in matter. *Physical Review*, **72**, 399–408.

Gershtein, S. S., Petrukhin, V. I., Ponomarev, L. I. & Prokoshkin, Yu. D. (1969). Meso-atomic processes and the model of large mesic molecules. *Soviet Physics Uspekhi*, **12**, 1–19.

Gershtein, S. S. & Ponomarev, L. I. (1975). Mesomolecular processes induced by μ^- and π^- mesons. In *Muon Physics*, vol. **3**, ed. V. W. Hughes & C. S. Wu, pp. 142–233. New York: Academic Press.

Goulard, B. & Primakoff, H. (1974). Nuclear muon-capture sum rules and mean nuclear excitation energies. *Physical Review*, **C10**, 2034–44.

Horvath, D. (1981). Chemistry of pionic hydrogen atoms. *Radiochimica Acta*, **28**, 241–54.

Ignatenko, A. E. (1961). Processes of depolarization of negative muons. *Nuclear Physics*, **23**, 75–89.

Kessler, D., Anderson, H. L., Dixit, M. S., Evans, H. J., McKee, R. J., Hargrove, C. K., Barton, R. D., Hincks, E. P. & McAndrew, J. D. (1967). μ^- atomic Lyman and Balmer series in Ti, TiO_2 and Mn. *Physical Review Letters*, **18**, 1179–83.

Knight, J. D., Orth, C. J., Schillaci, M. E., Naumann, R. A., Hartmann, F. J., Reidy, J. J. & Schneuwly, H. (1980). Coulomb capture ratios of negative muons in $N_2 + O_2$, NO and CO. *Physical Letters*, **79A**, 377–9.

Korenman, G. Ya. & Rogovaya, S. I. (1980). Primary populations in mesonic hydrogen atoms. *Journal of Physics*, **B13**, 641–53.

Lathrop, J. F., Lundy, R. A., Swanson, R. A., Telegdi, V. L. & Yovanovitch, D. D. (1960). Further experimental evidence concerning the Fermi–Teller Z-law. *Nuovo Cimento*, **15**, 831–4.

Leon, M. & Bethe, H. A. (1962). Negative meson absorption in liquid hydrogen. *Physical Review*, **127**, 636–47.

Nagamine, K. (1979). Negative muon spin rotation in solids. *Hyperfine Interactions*, **6**, 347–55.

Petrov, Yu. V. (1980). Muon catalysis for energy production by nuclear fusion. *Nature*, **285**, 466–8.

Ponomarev, L. I. (1973). Molecular structure effects on atomic and nuclear capture of mesons. *Annual Review of Nuclear Science*, **23**, 395–430.

Schneuwly, H., Dubler, T., Kaeser, K., Robert-Tissart, B., Schaller, L. A. & Schellenberg, L. (1978*a*). On the influence of the chemical bond on the relative muonic capture rates in elements of compounds. *Physics Letters*, **66A**, 188–90.

Schneuwly, H., Pokrovsky, V. N. & Ponomarev, L. I. (1978*b*). On coulomb capture ratios of negative mesons in chemical compounds. *Nuclear Physics*, **A312**, 419–26.

Sens, J. C., Swanson, R. A., Telegdi, V. L. & Yovanovitch, D. D. (1957). Experimental μ^--capture rates: evidence on exclusion principle effects and the type of interaction. *Physical Review*, **107**, 1464–5.

Sens, J. C., Swanson, R. A., Telegdi, V. L. & Yovanovitch, D. D. (1958). An experimental test of the Fermi–Teller Z-law. *Nuovo Cimento*, **7**, 536–44.

Skarsgard, L. D., Henkelman, R. M., Lam, G. K. Y. & Poon, M. N. (1979). Preclinical studies of the negative pi-meson beam at TRIUMF. *Radiation and Environmental Biophysics*, **16**, 193–204.

Suzuki, T. (1980). *A Systematic Study of Muon Capture*, Ph.D. Thesis, University of British Columbia, pp. 1–157.

Suzuki, T., Mikula, R. J., Garner, D. M., Fleming, D. G. & Measday, D. F. (1980). Muon capture in oxides using the lifetime method. *Physics Letters*, **95B**, 202–6.

Takagi, S., Yasuoka, H., Kuno, Y., Uemura, Y. J., Shibata, T. A., Hayano, R. S., Yamazaki, T., Ishikawa, Y., Kohn, S. E., Huang, C. Y. & Wernick, J. H. (1981). μ^-SR and doped Al-27 NMR studies of FeSi. *Hyperfine Interactions*, **8**, 499–502.

Vasilyev, V. A., Petrukhin, V. I., Risin, V. E., Suvorov, V. M. & Horvath, D. (1976). On the Z-dependence of the atomic capture rate of mesons in matter. *Joint Institute of Nuclear Research, USSR, Report P1-10222*, Dubra.

Vogel, P., Haff, P. K., Akylas, V. & Winther, A. (1974). Muon caputre in atoms, crystals and molecules. *Nuclear Physics*, **A254**, 445–79.

Wightman, A. S. (1950). Moderation of negative mesons in hydrogen. I: moderation from high energies to capture by an H_2 molecule. *Physical Review*, **77**, 521–8.

Yamazaki, T., Nagamiya, S., Hashimoto, O., Nagamine, K., Nakai, K., Sugimoto, K. & Crowe, K. M. (1974). Relativistic effect on magnetic moments of negative muons bound in high-Z nuclei. *Physics Letters*, **53B**, 117–20.

Zinov, V. G., Konin, A. D. & Mukhin, A. I. (1966). Atomic capture of negative muons in chemical compounds. *Soviet Journal of Nuclear Physics*, **2**, 613–18.

11

CONCLUDING CHAPTER

Perhaps the two most blatant omissions from the preceding chapters on μSR are background theory and solid-state applications. Justification for the dearth of theoretical derivations of the equations used in the text is offered on the grounds that each facet of μSR theory, from quantum electrodynamics to free radical hyperfine tensors, has already been dealt with elsewhere, better than could be accomplished here. Particularly commendable from the chemists' point of view are the treatments of Hughes (1966), Ivanter & Smilga (1968), Brewer, Crowe, Gygax & Schenck (1975), Schenck (1976), Percival & Fischer (1976), Brewer & Crowe (1978), Fleming *et al.* (1979), Garner (1979) and Roduner & Fischer (1981).

On the second major omission, it must be acknowledged that solid-state physics constitutes at least three-quarters of the total μSR research effort (Brewer & Crowe, 1978). This reflects the great value of the muon's unique magnetic moment, and its spin polarization, as a probe of structure and dynamics. The solid state encompasses more diversification than one should try to enumerate. Chemical compositions range from pure elements, through compounds and complex minerals, to an inexhaustible selection of 'doped' materials. Their magnetic properties include diamagnetics, paramagnetics, ferromagnetics and antiferromagnetics; electrical properties range from metals through semi-conductors to insulators; structures embrace ionic crystals, molecular crystals, amorphous phases and glassy media; and many of these properties change drastically with temperature, which is readily altered from several hundred degrees to almost zero. Mu and μ^+ (and μ^-) probe the local magnetic, nuclear and electronic structures, and they can mimic H and p^+ as interstitials. Thus one finds studies have been made of the microscopic magnetic environments, of anisotropies in field gradients, of dipolar fields, spin screening and its fluctuations, Knight shifts, phase transitions, lattice distortions, trapping and interstitial sites, diffusion, hopping and tunnelling, relaxation phenomena, anomalous muonium, indeed, muonium formation itself, and more. But most of this is the purview of physics: while our focus is on chemistry.

Nevertheless, some solid-state studies come under this aegis. For instance, the different effects occurring at the liquid–solid phase transition in water compared to ammonia and neopentane are important, and the effects in Ar, Kr and Xe represent a third variation (see Table 6.8). A solid of chemical interest is crystalline alum dodecahydrate ($AlK(SO_4)_2 \cdot 12H_2O$) in which the muonium yield was found to be the same as that in water, not ice (Brinkman & Roduner, unpublished data). Quartz is another solid which features prominently in chemical μSR. It was the first condensed phase medium in which muonium was observed (Myasishcheva, Obukhor, Roganov & Firsov, 1968), and in which the two-frequency precession due to breaking the ν_{12} and ν_{23} degeneracy in higher transverse fields was initially measured (Gurevich *et al.*, 1971). This splitting in quartz is a useful current tool for evaluating the frequency response of μSR equipment in the Mu-radical frequency range (see Figure 9.8). Fused and crystalline quartz seem to give the same values of P_M, with the implication that Mu formation and survival occur independently of trapping sites or molecular order. Crystalline quartz was the material in which a hyperfine quadrupolar precession of ^{T}Mu was found (Beder, Brewer & Spencer, 1979) following calculations on the magnitude of the quadrupole moment (Baryshevsky & Kuten, 1977; Beder, 1978). This effect is attributed to the slight D-wave character introduced by the hyperfine interaction in the triplet muonium ground state when there is a non-spherical electric field gradient, as is the case when a longitudinally polarized muon enters crystalline quartz perpendicular to the c-axis. Crystals of D and L quartz were also chosen for a test of the possible difference in interaction cross-section when polarized μ^+ form Mu with electrons present in molecules possessing different intrinsic helicities (Spencer, Fleming, Brewer & Mikula, 1979).

Solids show the full range of muon yields. Thus P_D varies from 0 to 1.0, having the latter value in most metals and some insulators. P_M ranges from zero in metals, through 0.7 in quartz to ~1.0 in solid Ar, Kr and Xe – much larger than in any liquid. P_R is also non-zero in unsaturated solids – in durene crystals for instance – and P_L varies almost to unity, though this may stem from short-lived ^{T}Mu.

The hyperfine splitting has been shown to be equal to the vacuum value in solid insulators such as ice, quartz, CO_2(s) and the noble-gas solids (Kiefl, Warren, Marshall & Oram, 1981); and in this respect they differ from the semiconductors Ge and Si, where electron screening apparently reduces the splitting by a factor of two.

In solid Xe the Mu spin relaxation rate is some ten times faster than in liquid

Xe: a result which has been attributed to the nuclear dipole moments of the isotopes ^{129}Xe and ^{131}Xe present at 26.4 and 21.2% respectively (Kiefl *et al.*, 1981). The greatly enhanced motional narrowing makes these nuclei much less effective in relaxing Mu in the liquid phase.

One of the original motivations in pursuing μSR studies was to apply fundamental tests of quantum electrodynamics (Hughes, 1966; Schenck, 1976). Because muons are leptons they are not subject to the strong interactions felt by hadrons, so very precise measurements of the muon's natural lifetime, its magnetic moment, and the muon-electron hyperfine structure interval in Mu, offer the most stringent tests of QED theory. In the same vein there is interest in determining the muonium-antimuonium conversion rate; but one needs Mu in a vacuum to study this because the rate is greatly reduced by the presence of matter with which Mu can interact (Marshall, 1981). Significant muonium fluxes *in vacuo* have been achieved by stopping muons in solids which have high surface area : volume ratios, such as very fine powders or thin foils. After being formed in the bulk solid, Mu can migrate to the surface where some escape into the evacuated space ready for study. Muonium atoms produced in this way have also been used to measure the Lamb shift ($2s_{1/2}$-$2p_{1/2}$ splitting) of the $n = 2$ level of Mu (Oram *et al.*, 1981) for which exact calculations are available (Owen, 1973). Furthermore, studies of Mu on powder surfaces have opened yet another area of research to which μSR can contribute, namely surface physics and chemistry (Kiefl, 1982).

What about future trends in muon chemistry itself? It is clear that Mu/H isotope effects are many and variable, but that a comprehensive picture is not yet available. Even when all the different kinetic isotope effects are resolved, one still has thermodynamic and spectroscopic aspects to consider, and here Mu-radical studies are invaluable. A consensus on the mechanism of Mu formation is needed, and the role of hot-atom reactions require further study. Many general properties of Mu and μ^+ are ripe for analysis, such as the standard reduction potential and acid dissociation constant of Mu, the solvation energy of μ^+ and its hydrogen-bonding properties. When these things are settled perhaps Mu will be registered as an isotope of H, and the IUPAC sequence rules for determining R or S and Z or E isomers will recognize its existence.

On the experimental front, one awaits the μSR resolution time reaching the point where chemical shifts in the ppm range can be measured, so that the different diamagnetic screenings of the muon's moment can be identified. This will mark a very important breakthrough, as mentioned in earlier chapters, enabling ready distinction between hot or thermal processes and permitting the diamagnetic muon's fate to be observed directly. At the other end of the time scale, the elimination of λ_0 as an artifact (if that is what it is) will allow slower reactions and relaxations to be studied. The low background of the pulsed μSR mode will be especially valuable in that regard. Also, the resonance and longitudinal-field relaxation methods (see Figure 3.1) may play more prominent roles in future, supplementing the transverse-field rotation method in chemical studies. There is also the possibility of utilizing a stroboscopic μSR method based on the accelerator's frequency (Schenck, 1976).

Finally, what about non-μSR prospects for muons? As far as muon and muonium chemistry are concerned it is difficult to envisage a monitoring technique which is more convenient, or more sensitive, or more insightful into the muon's behaviour than μSR is. But muon beams may be put to other uses because of their high degree of longitudinal polarization and particular *LET*. One such plan involves using muons to search for a connection between the unique handedness of the biosphere at the molecular level with the unique helicity of elementary particles produced in the weak interaction, as a possible origin of optical activity in nature. Another concerns the use of muons in single-particle pulse radiolysis based on *in situ* Cerenkov reabsorption spectroscopy; but both of these plans employ muons merely as polarized radiation sources, rather than solving the remaining problems in muon chemistry.

References for chapter 11

Baryshevsky, V. G. & Kuten, S. A. (1977). Quadrupole moment and quadrupole relaxation of spin of muonium and mesoatoms. *Physics Letters*, **64A**, 238–40.

Beder, D. (1978). The hyperfine quadrupole moment of muonium in the ground state. *Nuclear Physics*, **A305**, 411–17.

Beder, D. S., Brewer, J. H. & Spencer, D. P. (1979). Quadrupole splitting of muonium precession in α-quartz. *Physical Review Letters*, **42**, 808–11.

Brewer, J. H., Crowe, K. M., Gygax, F. N. & Schenck, A. (1975). Positive muons and muonium in matter. In *Muon Physics*, vol. **3**, ed. V. W. Hughes & C. S. Wu, pp. 3–139. New York: Academic Press.

Brewer, J. H. & Crowe, K. M. (1978). Advances in muon spin rotation. *Annual Reviews of Nuclear and Particle Science*, **28**, 239–326.

Fleming, D. G., Garner, D. M., Vaz, L. C., Walker, D. C., Brewer, J. H. & Crowe, K. M. (1979). Muonium chemistry – a review. *Advances in Chemistry Series*, **175**, 279–334.

Garner, D. M. (1979). *Application of the Muonium Spin Rotation Technique to a Study of the Gas Phase Chemical Kinetics of Muonium Reactions with the Halogens and Hydrogen Halides*. Ph.D. Thesis, University of British Columbia, pp. 1–307.

Gurevich, I. I., Ivanter, I. G., Meleshko, E. A., Nikolskii, B. A., Roganov, V. S., Selivanov, V. I., Smilga, V. P., Sokolov, B. V. & Shestakov, V. D. (1971). Two-frequency precession of muonium in a magnetic field. *Soviet Physics JETP*, **33**, 253–9.

Hughes, V. W. (1966). Muonium. *Annual Reviews of Nuclear Science*, **16**, 445–70.

Ivanter, I. G. & Smilga, V. P. (1968). Theory of the muonium mechanism of depolarization of μ^+ mesons in media. *Soviet Physics JETP*, **27**, 301–6.

Kiefl, R. F., Warren, J. B., Marshall, G. M. & Oram, C. J. (1981). Muonium in the condensed phases of Ar, Kr, and Xe. *Journal of Chemical Physics*, **74**, 308–13.

Kiefl, R. F. (1982). *Muonium and Positronium as Microprobes of Surfaces and Solids.* Ph.D. Thesis, University of British Columbia.

Marshall, G. M. (1981). *An Improved Upper Limit for Muonium Conversion to Antimuonium.* Ph.D. Thesis, University of British Columbia.

Myasishcheva, G. G., Obukhor, Yu. V., Roganov, V. S. & Firsov, V. G. (1968). A search for atomic muonium in chemically inert substances. *Soviet Physics, JETP*, **26**, 298–301.

Oram, C. J., Fry, C. A., Warren, J. B., Kiefl, R. F. & Brewer, J. H. (1981). Observation of the 2S state of muonium in a vacuum. *Journal of Physics*, **B14**, 789–93.

Owen, D. A. (1973). The muonium energy levels for $n = 2$. *Physics Letters*, **44B**, 199–201.

Percival, P. W. & Fisher, H. (1976). Theory and analysis of muon spin polarization in chemical systems. *Chemical Physics*, **16**, 89–99.

Roduner, E. & Fischer, H. (1981). Muonium substituted organic free radicals in liquid. Theory and analysis of μSR spectra. *Chemical Physics*, **54**, 261–76.

Schenck, A. (1976). On the application of polarized positive muons in solid state physics. In *Nuclear and Particle Physics at Intermediate Energies*, ed. J. B. Warren, pp. 159–297. New York: Plenum Press.

Spencer, D. P., Fleming, D. G., Brewer, J. H. & Mikula, R. J. (1979). A search for chirality-dependent muonium formation in quartz. In *Origins of Optical Activity in Nature*, ed. D. C. Walker, pp. 87–99. Amsterdam: Elsevier Scientific Publishing Co.

APPENDIX

For the convenience of quick consultation this appendix lists most of the current values for three quantities obtained by μSR: (A), the muon yields; (B), the muonium reaction rate constants; and (C), the hyperfine coupling constants of Mu-radicals. Neither statistical (fitting) error limits nor overall reproducibilities are given in these tables. Therefore, the reader must consult the original work in order to assess the accuracy of individual data. In many cases the possible errors are considerable, and in some cases mean values are recorded here. References are listed at the end of the three Tables.

(A) *Fractional yields of muon-containing species:* as observed in diamagnetic muon states (P_D), as free muonium atoms (P_M), in muonium-containing free radicals (P_R), and the remaining fraction (P_L). (A dash indicates that the yield does not seem to have been reported, nor can a reasonable inference about its value be drawn. Temperatures are given only when significantly different from ~295 K. Phases are given as gas (g), liquid (l), unspecified solid phase (s), metal (m) or crystal (c).)

Group	Substance	Phase (T(K))	P_D	P_M	P_R	P_L	Reference
Noble gases	He	g	1.0	0	0	0	*a*
		l (4)	>0.9	<0.02	0	<0.08	*b*
	Ne	g	1.0	0	0	0	*a*
	Ar	g	0.25	0.75	0	0	*c*
		l (85)	0.02	0.48	0	0.50	*d*
		s (77)	0.01	0.91	0	0.08	*d*
	Kr	g	0	1.0	0	0	*a*
		l (120)	0.07	0.57	0	0.36	*d*
		s (90)	0.01	0.71	0	0.28	*d*
	Xe	g	0	1.0	0	0	*a*
		l (162)	0.03	0.43	0	0.54	*d*
		s (150)	0.05	0.79	0	0.16	*d*

Group	Substance	Phase (T (K))	p_D	p_M	p_R	p_L	Reference
Gases at STP	H_2	g	0.4	0.6	0	0	*e*
	N_2	g	0.16	0.84	0	0	*e*
	O_2	g	0.44	–	–	–	*f*
	N_2O	g	0.10	–	–	–	*f*
	SF_6	g	0.75	–	–	–	*f*
	CH_4	g	0.12	0.88	0	0	*e*
	NH_3	l (210)	–	0.21	–	–	*g*
		s (200)	–	0.21	–	–	*g*
	$C(CH_3)_4$	l	0.55	0.18	0	0.27	*h*
		s (209)	0.61	0.19	0	0.20	*h*
Liquids at STP (non-aromatic)	H_2O	g (420)	0.07	0.93	0	0	*e*
		l	0.62	0.20	0	0.18	*i*
		s (270)	0.48	0.52	0	0	*i*
	D_2O	l	0.57	0.18	0	0.25	*i*
		s (270)	0.39	0.63	0	0	*i*
	CH_3OH	g (420)	0.13	0.87	0	0	*e*
		l	0.61	0.19	0	0.20	*j*
	CD_3OH	l	0.51	0.31	0	0.18	*j*
	C_2H_5OH	l	0.59	0.20	0	0.21	*j*
	$(CH_3)_2CHOH$	l	0.62	–	–	–	*k*
	glycerol	l	0.75	–	–	–	*k*
	$(CHOH)_6$	s	0.7	0	–	–	*l*
	$(CH_3)_2CO$	l	0.54	–	>0	–	*k, m*
	$Si(CH_3)_4$	l	0.53	0.21	0	0.26	*n*
	n-hexane	g	0.19	0.81	0	0	*e*
		l	0.65	0.13	0	0.22	*n*
	n-hexene	l	0.50	–	–	–	*k*
	n-hexyne	l	0.43	–	–	–	*k*
	c-hexane	g	0.17	0.83	0	0	*e*
		l	0.69	0.20	0	0.11	*n*
	c-hexadiene-1,4	l	0.40	–	–	–	*k*
	c-hexadiene-1,3	l	0.32	–	–	–	*k*
	CS_2	l	0.16	0	?	–	*k*
	CCl_4	g	0.5	0	0.5	0	*e*
		l	1.0	0	0	0	*k*
	$CHCl_3$	g	0	1.0	0	0	*e*
		l	0.85	–	–	–	*k*
	CH_2Cl_2	g	0	1.0	0	0	*e*
		l	0.70	–	–	–	*k*
	$SiCl_4$	l	0.48	–	–	–	*k*
	$SnCl_4$	l	0.99	–	–	–	*k*
	$TiCl_4$	l	1.0	–	–	–	*k*
	$CHBr_3$	l	0.85	–	–	–	*p*
	CH_3CN	l	0.43	0	>0	–	*l*
	$CH_2{=}CHCN$	l	0.28	0	>0	–	*l*
	$CH_2{=}CHCO_2H$	l	0.51	0	>0	–	*l*
	$CH_2{=}CHCO_2C_2H_5$	l	0.41	0	>0	–	*l*
	$CH_2{=}C(CH_3)CO_2$	l	0.38	0	>0	–	*l*
	$CH_2{=}C(CH_3)CO_2C_2$	l	0.38	0	>0	–	*l*

Group	Substance	Phase (T(K))	P_D	P_M	P_R	P_L	Reference
Aromatic liquids	C_6H_6	1	0.15	0	0.65	0.20	*q*
	C_6H_5F	1	0.19	0	0.43	0.38	*q*
	$C_6H_4F_2(\bar{3})$	1	0.25	0	0.45	0.3	*q*
	$C_6H_3F_3$	1	0.25	0	0.3	0.45	*q*
	$C_6H_2F_4(\bar{3})$	1	0.25	0	0.35	0.4	*q*
	C_6HF_5	1	0.23	0	0.14	0.63	*q*
	C_6F_6	1	0.20	0	0.35	0.45	*q*
	$C_6H_5CH_3$	1	0.25	0	0.50	0.25	*q*
	$C_6H_4(CH_3)_2(\bar{3})$	1	0.33	0	0.4	0.25	*q*
	$C_6H_3(CH_3)_3(\bar{3})$	1	0.34	0	0.4	0.25	*q*
	C_6H_5Cl	1	0.23	–	–	–	*k*
	C_6H_5Br	1	0.38	–	–	–	*k*
	C_6H_5I	1	0.49	–	–	–	*k*
	$C_6H_5CH_2Cl$	1	0.35	–	–	–	*k*
	$C_6H_5CHCl_2$	1	0.46	–	–	–	*k*
	$C_6H_5CCl_3$	1	0.61	–	–	–	*k, l*
	C_6H_5OH	1	0.38	–	–	–	*k*
	$CH_2{=}CHC_6H_5$	1	0.17	0	>0	–	*l*
Solids at STP	Al	m	1.0	0	0	0	*r*
	quartz	s, c	0.21	0.7	0	0.1	*r, l*
	graphite	c	1.0	0	0	0	*s*
	diamond	c	0.2	–	–	–	*s*
	polyethylene	s	0.64	–	–	–	*s*
	polystyrene	s	0.31	–	–	–	*s*
	polymethyl-methacrylate	s	0.38	–	–	–	*l*
	teflon	s	0.80	–	–	–	*p*
	P	s	0.11	–	–	–	*s*
	S	s	0.06	–	–	–	*s*
	CsI	c	0.14	–	–	–	*s*
	NaCl	c	0.18	–	–	–	*s*
	MgF_2	c	0.59	–	–	–	*s*
	MgO	c	0.35	–	–	–	*s*
	Dry Ice (CO_2)	s	0.2	–	–	–	*p*

For other P_D values in various materials see *p*, *r* and *s*, in particular.

(B) *Muonium reaction rate constants:* i.e. k_M for reaction of Mu with the solute named, at ~295 K

Solute	k_M ($M^{-1} s^{-1}$)	Reference
(i) *In dilute aqueous solution*		
(a) Inorganic solutes: alphabetical in chemical symbol		
Ag^+	1.6×10^{10}	*j*
Cd^{2+}	8×10^6	*t*
$Cd(CN)_4^{2+}$	1.7×10^{10}	*l*
ClO_4^-	10^7	*u*
CN^-	3×10^9	*l*
CNS^-	6×10^7	*v*
Cr^{3+}	5.3×10^9	*u*
CrO_4^{2-}	2.4×10^{10}	*w*
Cu^{2+}	6.5×10^9	*u*
Fe^{2+}	1.2×10^{10}	*u*
$Fe(CN)_6^{4-}$	3.0×10^8	*u*
Fe^{3+}	5.5×10^9	*u*
$Fe(CN)_6^{3-}$	2.0×10^{10}	*u*
$HgCl_2$	2×10^9	*u*
I^-	7×10^7	*x*
I_2	1.7×10^{10}	*x*
I_3^-	1×10^9	*l*
MnO_4^-	2.5×10^{10}	*j*, *z*
Ni^{2+}	1.7×10^{10}	*u*, *z*
$Ni(NH_3)_4^{2+}$	1.5×10^{10}	*l*
$Ni(cyclam)^{2+}$	5×10^8	*l*
$Ni(cyclam)(NH_3)_2^{2+}$	2×10^{10}	*l*
NO_3^-	1.5×10^9	*j*, *z*
OH^-	1.7×10^7	*j*
O_2	2.4×10^{10}	*y*
$S_2O_3^{2-}$	1.4×10^{10}	*g*
Tl^+	8×10^8	*v*
Zn^{2+}	$<10^7$	*v*
$K^+/SO_4^{2-}/NH_4^+$	$<10^7$	*v*
$Na^+/H^+/Cl^-$	$<2 \times 10^5$	*j*
(b) Organic solutes: alphabetic in common or systematic name		
acetone	8.7×10^7	*j*
acetonitrile	5.1×10^7	*l*
acrylamide	1.9×10^{10}	*l*
acrylic acid	1.5×10^{10}	*l*
acrylonitrile	1.1×10^{10}	*l*
ascorbic acid	1.8×10^9	*j*
bromoacetic acid	1.5×10^9	*j*
2-bromopropionic acid	4×10^9	*j*
3-bromopropionic acid	$\sim 3 \times 10^8$	*j*
2-butanol	1.1×10^6	*j*
cyanoacetate	7.7×10^7	*l*

cyclodextrin (α or β)	$\sim 2 \times 10^7$	x
DNA and its nucleotides	$(4\text{–}9) \times 10^9$	aa
dihydroxyfumaric acid	4.5×10^7	j
ethanol	$<3 \times 10^5$	j
formate ion	8×10^6	j, z
formate ion–d (DCO_2^-)	10^6	bb
fumaric acid	1.4×10^{10}	j
haemin	2.7×10^9	cc
maleic acid	1.1×10^{10}	j, z
methanol	3×10^4	n
methylmethacrylate	9.5×10^9	l
naphthalene	1.3×10^9	dd
p-nitrophenol	8×10^9	v
phenol	7×10^9	v
2-propanol	5×10^5	j, t
protoporphyrin	6.2×10^8	cc
sodium hexyl sulphate	$<10^7$	dd
styrene	1.1×10^9	l

(ii) *Gas phase reactions* (Ar or N_2 as moderators)
A complete list is given as Table 7.1, page 101.

(C) *Hyperfine coupling constants for Mu-radicals*, in liquids at ~295 K (unless otherwise specified)

Type	Parent molecule	Radical	A_μ (MHz)	Reference
Terminal alkenes	ethene	$CH_2Mu\dot{C}H_2$	331.0	ee
	propene	$CH_2Mu\dot{C}HCH_3$	303.7	ee
	2-methylpropene	$CH_2Mu\dot{C}(CH_3)_2$	291.6	ee
	pentene-1	$CH_2Mu\dot{C}HCH_2CH_2CH_3$	306.9	ee
	3,3-dimethylbutene-1	$CH_2Mu\dot{C}HC(CH_3)_3$	311.0	ee
	2,4,4-trimethylpentene-1	$CH_2Mu\dot{C}(CH_3)CH_2C(CH_3)_3$‡	289.1	ee
	isopropylenecyclopropane	$CH_2Mu\dot{C}(CH_3)C_3H_5$	281.4	ee
	heptadiene-1,6	$CH_2Mu\dot{C}H(CH_2)_3CH{=}CH_2$	307.5	ee
	1-butene-4 ol	$CH_2Mu\dot{C}HCH_2CH_2OH$	308.5	ee
	1-butene-3 ol	$CH_2Mu\dot{C}HCH(OH)CH_3$	311.0	ee
	allylethylether	$CH_2Mu\dot{C}HCH_2OCH_2CH_3$	309.7	ee
	allylpropylether	$\dot{C}H_2CHMuCH_2OC_3H_7$	336.1	ee
		$CH_2Mu\dot{C}HCH_2OC_3H_7$	315.5	ee
	diallylether	$\dot{C}H_2CHMuCH_2OCH_2CH{=}CH_2$	326.9	ee
		$CH_2Mu\dot{C}HCH_2OCH_2CH{=}CH_2$	309.7	ee
	styrene	$CH_2Mu\dot{C}HC_6H_5$	213.5	l, ee
	acrylonitrile	$CH_2Mu\dot{C}HCN$	266.5	ee
	ethylacrylate	$CH_2Mu\dot{C}HCO_2C_2H_5$	317.0	ee (253 K)
	methylmethacrylate	$CH_2Mu\dot{C}(CH_3)CO_2CH_3$	279	l
	ethylmethacrylate	$CH_2Mu\dot{C}(CH_3)CO_2C_2H_5$	274.8	ee
		(cis and trans isomers)	269.7	ee

Type	Parent molecule	Radical	A_μ (MHz)	Reference
	vinylacetate	$CH_2Mu\dot{C}HOCOCH_3$	276.0	*ee*
Non-terminal alkenes	pentene-2	$CH_3\dot{C}HCHMuCH_2CH_3$	297.0	*ee*
		$CH_3CHMu\dot{C}HCH_2CH_3$	305.0	*ee*
	2-methylbutene-2	$(CH_3)_2\dot{C}CHMuCH_3$	161.0	*ee*
	2,3-dimethylbutene-2	$(CH_3)_2CMu\dot{C}(CH_3)_2$	145.6	*ee*, *ff*
	2,4,4-trimethylpentene-2	$(CH_3)_2CMu\dot{C}HC(CH_3)_3$	251.2	*ee*
	2-butene-4-ol	$CH_3\dot{C}HCHMuCH_2OH$	287.5	*ee*
		$CH_3CHMu\dot{C}HCH_2OH$	306.6	*ee*
	2-butene-1,4-diol	$HOCH_2CHMu\dot{C}HCH_2OH$	284.9	*ee* (288 K)
Congugated dienes	butadiene		188.3	*ee*
	pentadiene-1,3		182.5	*ee*
	4-methylpentadiene-1,3		173.9	*ee*
	2-methylbutadiene		180.6	*ee*
	2-methylpentadiene-1,3		176.2	*ee*
	2,4-dimethylpentadiene-1,3		166.6	*ee*
	2,4-dimethylpentadiene-1,3		158.7	*ee*
	2-methylbutadiene		199.4	*ee*
	2,3-dimethylbutadiene		196.2	*ee*
	pentadiene-1,3		168.9	*ee*
	hexadiene-2,4		162.8	*ee*
	2-methylpentadiene-1,3		187.3	*ee*
	2,5-dimethylhexadiene-2,4		114.2	*ee*
	hexadiene-2,4		291.6	*ee*
	2,5-dimethylhexadiene-2,4		251.2	*ee*
Aromatics		Radicals are all substituted cyclohexadienyl radicals. With Mu at C1 the CH_3 or F occupies these positions in the radical:		
	benzene	(C_6H_6Mu)	514.6	*l*, *q*, *m*

Type	Parent molecule	Radical	A_μ (MHz)	Reference
	benzene-d_6	(C_6D_6Mu)	520.5	*q*, *m*
	toluene	2	489.6	*q*
		4	496.4	*q*
		3	509.3	*q*
	1,4-dimethylbenzene	2,5	482.8	*q*
	1,2-dimethylbenzene	2,3	493.5	*q*
		3,4	496.2	*q*
	1,3-dimethylbenzene	2,6	466.4	*q*
		2,4	474.4	*q*
		1,3	484.6	*q*
		3,5	504.0	*q*
	1,3,5-trimethylbenzene	2,4,6	453.1	*q*
		1,3,5	483.9	*q*
	1,2,3-trimethylbenzene	1,2,3 or 1,2,6‡	418.8	*q*
		2,3,4	481.8	*q*
		3,4,5	492.7	*q*
	1,2,4-trimethylbenzene	2,3,6	468.1	*q*
		2,4,5	472.1	*q*
		2,3,5	486.4	*q*
	1,2,4,5-tetramethyl-benzene(*c*)	Anisotropic	454.2–491.4	*gg*
	fluorobenzene	2	485.7	*q*
		3 or 4	511.8	*q*
	1,4-difluorobenzene	2,5	477.7	*q*
	1,2-difluorobenzene	2,3	491.0	*q*
		3,4	501.4	*q*
	1,3-difluorobenzene	2,6	467.4	*q*
		2,4	486.5	*q*
		3,5	513.7	*q*
	1,2,4-trifluorobenzene	2,3,6	466.0	*q*
		2,4,5	472.1	*q*
		2,3,5	486.8	*q*
	1,2,3,4-tetrafluoro-benzene	2,3,4,5	471.5	*q*
		ipso‡	230.2	*q*
	1,2,3,5-tetrafluoro-benzene	2,3,4,6	464.1	*q*
		ipso‡	260.1	*q*
	1,2,4,5-tetrafluoro-benzene	2,3,5,6	468.3	*q*
		1,2,4,5	234.4	*q*
	pentafluorobenzene	2–6	453.4	*q*
	hexafluorobenzene	1–6	200.9	*q*
	ααα-trifluorotoluene	1	474	*gg*
		2	500.3	*gg*
		4	508.7	*gg*
		3	510.5	*gg*
	thiophene	2-thiophenyl	338.7	*m*
	deuterated toluenes		490–511	*q*
Carbonyl	acetone	$(CH_3)_2\dot{C}OMu$	26.1	*m*

Errors in A_μ in the range 0.2–0.6 MHz.
* Denotes position of Mu.
‡ Assignment uncertain.
(*c*) Single crystal.

References for Appendix

a Stambaugh, R. D., Casperson, D. E., Crane, T. W., Hughes, V. W., Kaspar, H. F., Sonder, P., Thompson, P. A., Orth, H., Putlitz, G. zu. & Denison, A. B. (1974). Muonium formation in noble gases and noble-gas mixtures. *Physical Review Letters*, **33**, 568–71.

b Crane, T. W., Casperson, D. E., Chang, H., Hughes, V. W., Kaspar, H. F., Lovett, B., Schiz, A., Sonder, P., Stambaugh, R. D. & Putlitz, G. zu. (1974). Behaviour of positive muons in liquid helium. *Physical Review Letters*, **33**, 572–4.

c Barnett, B. A., Chang, C. Y., Yodh, G. B., Carroll, J. B., Eckhause, M., Heieh, C. S., Kane, J. R. & Spence, C. B. (1975). Muonium-formation measurement in low pressure argon gas. *Physical Review A*, **11**, 39–41.

d Keifl, R. F., Warren, J. B., Marshall, G. M. & Oram, C. J. (1981). Muonium in the condensed phases of Ar, Kr and Xe. *Journal of Chemical Physics*, **74**, 308–13.

e Fleming, D. G. (1981). *Annual Report, TRIUMF*, pp. 39–43; Arseneau, D. *et al.* (1982), to be published.

f Hughes, V. W., McColm, D. W., Ziock, K. & Prepost, R. (1970). Muonium I. Muonium formation and Larmor precession. *Physical Review A*, **1**, 595–617.

g Percival, P. W. (1981). *Annual Report, TRIUMF*, pp. 43–5.

h Ng, B. W., Stadlbauer, J. M., Jean, Y. C. & Walker, D. C. (1982). Muonium atoms in liquid and solid neopentane. *Canadian Journal of Chemistry* (in press).

i Percival, P. W., Roduner, E. & Fischer, H. (1978). Radiolysis effects in muonium chemistry. *Chemical Physics*, **32**, 353–67.

j Percival, P. W., Roduner, E. & Fischer, H. (1979). Radiation chemistry and reaction kinetics of muonium in liquids. *Advances in Chemistry Series*, **175**, 335–55.

k Fleming, D. G., Garner, D. M., Vaz, L. C., Walker, D. C., Brewer, J. H. & Crowe, K. M. (1979). Muonium chemistry – a review. *Advances in Chemistry Series*, **175**, 279–334.

l (i) Stadlbauer, J. M., Ng, B. W., Walker, D. C., Jean, Y. C. & Ito, Y. (1981). Muonium addition to vinyl monomers. *Canadian Journal of Chemistry*, **59**, 3261–6. (ii) Stadlbauer, J. M., Ng, B. W., Jean, Y. C. & Walker, D. C. (1982). Spin conversion reaction of muonium with nickel cyclam. *Journal of the American Chemical Society* (in press). (iii) Stadlbauer, J. M., Ng, B. W., Jean, Y. C. & Walker, D. C. (1982). Muonium addition to cyanides (submitted for publication). (iv) Stadlbauer, J. M. *et al.* (1982), unpublished data.

m Roduner, E. (1979). *On the Liquid Phase Chemistry of the Light Hydrogen Isotope Muonium*. Ph.D. Thesis, University of Zurich, pp. 1–97.

n Ito, Y., Ng, B. W., Jean, Y. C. & Walker, D. C. (1980). Muonium atoms observed in liquid hydrocarbons. *Canadian Journal of Chemistry*, **58**, 2395–401.

p Myasishcheva, G. G., Obukhov, Yu. V., Roganov, V. S. & Firsov, V. G. (1967). The chemistry of muonium. *High Energy Chemistry*, **1**, 389–93; **3**, 510–14.

q Roduner, E., Brinkman, G. A. & Louwrier, W. F. (1982). Muonium substituted organic free radicals in liquids. Muon-electron hyperfine coupling constants and the selectivity of formation of methyl and fluorine substituted cyclohexadienyl type radicals. *Chemical Physics* (submitted).

r Brewer, J. H., Crowe, K. M., Gygax, F. N. & Schenck, A. (1975). Positive muons and muonium in matter. In *Muon Physics*, vol. **3**, ed. V. W. Hughes & C. S. Wu, pp. 3–139. New York: Academic Press.

s Swanson, R. A. (1958). Depolarization of positive muons in condensed matter. *Physical Review*, **112**, 580–6.

t Walker, D. C., Jean, Y. C. & Fleming, D. G. (1979). Muonium atoms and intraspur processes in water. *Journal of Chemical Physics*, **70**, 4534–41.
u Jean, Y. C., Brewer, J. H., Fleming, D. G. & Walker, D. C. (1978). Spin-conversion of muonium by interaction with paramagnetic ions. *Chemical Physics Letters*, **60**, 125–9.
v Jean, Y. C., Brewer, J. H., Fleming, D. G., Garner, D. M., Mikula, R. J., Vaz, L. C. & Walker, D. C. (1978). Reactivity of muonium atoms in aqueous solution. *Chemical Physics Letters*, **57**, 293–7.
w Percival, P. W. (1981). Muonium formation in water and aqueous solutions. *Hyperfine Interactions*, **8**, 315–23.
x Jean, Y. C., Ng, B. W., Ito, Y., Nguyen, T. Q. & Walker, D. C. (1981). MSR applications to muonium reactivity in cyclodextrins. *Hyperfine Interactions*, **8**, 351–4.
y Jean, Y. C., Fleming, D. G., Ng, B. W. & Walker, D. C. (1979). Reaction of muonium with O_2 in aqueous solution. *Chemical Physics Letters*, **66**, 187–90.
z Ng, B. W., Jean, Y. C., Ito, Y., Suzuki, T., Brewer, J. H., Fleming, D. G. & Walker, D. C. (1981). Diffusion and activatin-controlled reactions of muonium in aqueous solutions. *Journal of Physical Chemistry*, **85**, 454–8.
aa Bucci, C., Guidi, G., De'munari, G. M., Manfredi, M., Podine, P., Tedeschi, R., Crippa, P. R. & Vecli, A. (1979). Interaction of muonium with molecules of biological interest in water solution. *Hyperfine Interaction*, **6**, 425–9.
bb Ng, B. W. *et al.* (1982), to be published.
cc Jean, Y. C., Ng, B. W. & Walker, D. C. (1980). Chemical reactions between muonium and porphyrins. *Chemical Physics Letters*, **75**, 561–4.
dd Jean, Y. C., Ng. B. W., Stadlbauer, J. M. & Walker, D. C. (1981). Muonium reactions in micelles. *Journal of Chemical Physics*, **75**, 561–4.
ee Roduner, E., Strub, W., Burkhard, P., Hochmann, J., Percival, P. W., Fischer, H., Ramos, M. & Webster, B. C. (1982). Muonium substituted organic free radicals in liquids. Muon-electron hyperfine coupling constants of alkyl and allyl radicals. *Chemical Physics*, **67**, 275–85.
ff Roduner, E., Percival, P. W., Fleming, D. G., Hochman, J. & Fischer, H. (1978). Muonium-substituted transient radicals observed by muon spin rotation. *Chemical Physics Letters*, **57**, 37–40.
gg Roduner, E. (1981). Observation of muonium substituted free radicals in a durene single crystal. *Chemical Physics Letters*, **81**, 191–4.

INDEX